Discover And Contact

By

Ian Beardsley

Copyright © 2010-2014 by Ian Beardsley

ISBN:978-1-312-08470-4

All That Can Be Said

They originated in the Far East, and passed some through the south, and others through the north over a period of A Thousand Years through deserts. They became Artists, and camped where there was no one else, and went unseen and unknown during all that time, only to unite and settle at Land's End, in neighborhoods, barrios, or quarters of western construct, and to let out only one verse of one of their poets, who is anonymous: "If there is someone in the street, he is familiar with it. If there is someone in the street, he knows him."

Chapter 1

AE-35

I wrote a short story last night, called Gypsy Shamanism and the Universe about the AE-35 unit, which is the unit in the movie and book 2001: A Space Odyssey that HAL reports will fail and discontinue communication to Earth. I decided to read the passage dealing with the event in 2001 and HAL, the ship computer, reports it will fail in within 72 hours. Strange, because Venus is the source of 7.2 in my Neptune equation and represents failure, where Mars represents success.

Ian Beardsley
August 5, 2012

It must have been 1989 or 1990 when I took a leave of absence from The University Of Oregon, studying Spanish, Physics, and working at the state observatory in Oregon -- Pine Mountain Observatory—to pursue flamenco in Spain.

The Moors, who carved caves into the hills for residence when they were building the Alhambra Castle on the hill facing them, abandoned them before the Gypsies, or Roma, had arrived there in Granada Spain. The Gypsies were resourceful enough to stucco and tile the abandoned caves, and take them up for homes.

Living in one such cave owned by a gypsy shaman, was really not a down and out situation, as these homes had plumbing and gas cooking units that ran off bottles of propane. It was really comparable to living in a Native American adobe home in New Mexico.

Of course living in such a place came with responsibilities, and that included watering its gardens. The Shaman told me: "Water the flowers, and, when you are done, roll up the hose and put it in the cave, or it will get stolen". I had studied Castilian Spanish in college and as such a hose is "una manguera", but the Shaman called it "una goma" and goma translates as rubber. Roll up the hose and put it away when you are done with it: good advice!

So, I water the flowers, rollup the hose and put it away. The Shaman comes to the cave the next day and tells me I didn't roll up the hose and put it away, so it got stolen, and that I had to buy him a new one.

He comes by the cave a few days later, wakes me up asks me to accompany him out of The Sacromonte, to some place between there and the old Arabic city, Albaicin, to buy him a new hose.

It wasn't a far walk at all, the equivalent of a few city blocks from the caves. We get to the store, which was a counter facing the street, not one that you could enter. He says to the man behind the counter, give me 5 meters of hose. The man behind the counter pulled off five meters of hose from the spindle, and cut the hose to that length. He stated a value in pesetas, maybe 800, or so, (about eight dollars at the time) and the Shaman told me to give that amount to the man behind the counter, who was Spanish. I paid the man, and we left.

I carried the hose, and the Shaman walked along side me until we arrived at his cave where I was staying. We entered the cave stopped at the walk way between living room and kitchen, and he said: "follow me". We went through a tunnel that had about three chambers in the cave, and entered one on our right as we were heading in, and we stopped and before me was a collection of what I estimated to be fifteen rubber hoses sitting on ground. The Shaman told me to set the one I had just bought him on the floor with the others. I did, and we left the chamber, and he left the cave, and I retreated to a couch in the cave living room.

Chapter 2

Gypsies have a way of knowing things about a person, whether or not one discloses it to them in words, and The Shaman was aware that I not only worked in Astronomy, but that my work in astronomy involved knowing and doing electronics.

So, maybe a week or two after I had bought him a hose, he came to his cave where I was staying, and asked me if I would be able to install an antenna for television at an apartment where his nephew lived.

So this time I was not carrying a hose through The Sacromonte, but an antenna.

There were several of us on the patio, on a hill adjacent to the apartment of The Shaman's Nephew, installing an antenna for television reception.

Chapter 3

I am now in Southern California, at the house of my mother, it is late at night, she is a asleep, and I am about 24 years old and I decide to look out the window, east, across The Atlantic, to Spain. Immediately I see the Shaman, in his living room, where I had eaten a bowl of the Gypsy soup called Puchero, and I hear the word Antenna. I now realize when I installed the antenna, I had become one, and was receiving messages from the Shaman.

The Shaman's Children were flamenco guitarists, and I learned from them, to play the guitar. I am now playing flamenco, with instructions from the shaman to put the gypsy space program into my music. I realize I am not just any antenna, but the AE35 that malfunctioned aboard The Discovery just before it arrived at the planet Jupiter in Arthur C. Clarke's and Stanley Kubrick's "2001: A Space Odyssey". The Shaman tells me, telepathically, that this time the mission won't fail.

Chapter 4

I am watching Star Wars and see a spaceship, which is two oblong capsules flying connected in tandem. The Gypsy Shaman says to me telepathically: "Dios es una idea: son dos". I understand that to mean "God is an idea: there are two elements". So I go through life basing my life on the number two.

Chapter 5

Once one has tasted Spain, that person longs to return. I land in Madrid, Northern Spain, The Capitol. The Spaniards know my destination is Granada, Southern Spain, The Gypsy Neighborhood called The Sacromonte, the caves, and immediately recognize I am under the spell of a Gypsy Shaman, and what is more that I am The AE35 Antenna for The Gypsy Space Program. Flamenco being flamenco, the Spaniards do not undo the spell, but reprogram the instructions for me, the AE35 Antenna, so that when I arrive back in the United States, my flamenco will now state their idea of a space program. It was of course, flamenco being flamenco, an attempt to out-do the Gypsy space program.

Chapter 6

I am back in the United States and I am at the house of my mother, it is night time again, she is asleep, and I look out the window east, across the Atlantic, to Spain, and this time I do not see the living room of the gypsy shaman, but the streets of Madrid at night, and all the people, and the word Jupiter comes to mind and I am about to say of course, Jupiter, and The Spanish interrupt and say "Yes, you are right it is the largest planet in the solar system, you are right to consider it, all else will flow from it."

I know ratios, in mathematics are the most interesting subject, like pi, the ratio of the circumference of a circle to its diameter, and the golden ratio, so I consider the ratio of the orbit of Saturn (the second largest planet in the solar system) to the orbit of Jupiter at their closest approaches to The Sun, and find it is nine-fifths (nine compared to five) which divided out is one point eight (1.8).

I then proceed to the next logical step: not ratios, but proportions. A ratio is this compared to that, but a proportion is this is to that as this is to that. So the question is: Saturn is to Jupiter as what is to what? Of course the answer is as Gold is to Silver. Gold is divine; silver is next down on the list. Of course one does not compare a dozen oranges to a half dozen apples, but a dozen of one to a dozen of the other, if one wants to extract any kind of meaning. But atoms of gold and silver are not measured in dozens, but in moles. So I compared a mole of gold to a mole of silver, and I said no way, it is nine-fifths, and Saturn is indeed to Jupiter as Gold is to Silver.

I said to myself: How far does this go? The Shaman's son once told me he was in love with the moon. So I compared the radius of the sun, the distance from its center to its surface to the lunar orbital radius, the distance from the center of the earth to the center of the moon. It was Nine compared to Five again!

Chapter 7

I had found 9/5 was at the crux of the Universe, but for every yin there had to be a yang. Nine fifths was one and eight-tenths of the way around a circle. The one took you back to the beginning which left you with 8 tenths. Now go to eight tenths in the other direction, it is 72 degrees of the 360 degrees in a circle. That is the separation between petals on a five-petaled flower, a most popular arrangement. Indeed life is known to have five-fold symmetry, the physical, like snowflakes, six-fold. Do the algorithm of five-fold symmetry in reverse for six-fold symmetry, and you get the yang to the yin of nine-fifths is five-thirds.

Nine-fifths was in the elements gold to silver, Saturn to Jupiter, Sun to moon. Where was five-thirds? Salt of course. "The Salt Of The Earth" is that which is good, just read Shakespeare's "King Lear". Sodium is the metal component to table salt, Potassium is, aside from being an important fertilizer, the substitute for Sodium, as a metal component to make salt substitute. The molar mass of potassium to sodium is five to three, the yang to the yin of nine-fifths, which is gold to silver. But multiply yin with yang, that is nine-fifths with five-thirds, and you get 3, and the earth is the third planet from the sun.

I thought the crux of the universe must be the difference between nine-fifths and five-thirds. I subtracted the two and got two-fifteenths! Two compared to fifteen! I had bought the Shaman his fifteenth rubber hose, and after he made me into the AE35 Antenna one of his first transmissions to me was: "God Is An Idea: There Are Two Elements".

It is so obvious, the most abundant gas in the Earth Atmosphere is Nitrogen, chemical symbol 15!

Chapter Eight: A Witch Made The Universe

First the Gypsy Shaman, Manuel, introduces the idea of Taoism into the science of the Universe, saying that the Universe is composed of two opposite, but complementary forces, the yin and yang of Taoism when he says that God is an idea, there are two elements (Dios es una idea, son dos). When he says God, we know he means it in the Einsteinian sense of God means Nature.

We later learn that yin is nine-fifths and yang is five-thirds by looking at Nature. Then we derive from them the Yin and Yang equations. After that we find we are lacking a value over which to integrate them. The Shaman solves that by introducing the number 15, which is the number of rubber hoses in his rubber hose collection he keeps in an ancient Moorish cave in Spain.

We only require his counterpart, but once lured in, he makes him into the AE-35 Antenna that was aboard the spacecraft Discovery in the movie 2001: A Space Odyssey so they can communicate with one another telepathically across the Atlantic.

But the extraterrestrials subvert the plan by making a transmission to Earth on August 15, 1977 that was received by SETI that lasted 72 seconds, the same 72 that was the number of hours within which the AE-35 antenna would fail as reported by Discovery Ship Computer, HAL to crewmen Dave Bowman and Frank Poole.

Once Manuel's counterpart realizes the 72 of the duration of the SETI signal from extraterrestrials is not only the same as the 72 of the hours for within which HAL reports failure of the AE-35 Antenna and what is more realizes that the planet Venus is 0.72 Astronomical Units from the Sun, and is considered a failed Earth by Astronomers, that there is now failure clear across the board, and, the Gypsy's spell is broken, and the extraterrestrials successfully sever the telepathic communication between Shaman and his counterpart.

The counterpart now thinking on his own begins to wonder why the extraterrestrials felt they needed to sever his connection with the Gypsy Shaman, and whether the AE-35 antenna has a deeper meaning, and becomes very curious about it. He starts playing the Indian Tabla to try and learn the secrets of AE-35 telepathy.

He felt the place to look was in a transmission he received from the Shaman's mother, and grandmother of his children, who taught him the art of Deep Song on the guitar, which was that a witch made the universe. He still had the discoveries about the universe of the yin and yang equations, and was convinced that the Gypsy Shaman and he were making great progress in solving a riddle that ran very deep.

Chapter Nine

Let us look at what we know: God is an idea, there are two elements and a witch made the Universe. Universe, Nature, God – these are all interchangeable. We might as well say The Universe is an idea, there are two elements and a witch made the Universe. But we have not really looked at the first part of the statement; the universe is an idea. If it is an idea it is not something real, it is a dream in the mind of the beholder. We have discovered that the two elements that make the idea are nine-fifths and five-thirds. The idea is then, their product. The product of nine-fifths and five-thirds is three, and three points are the fewest that enclose an area, which takes the form of a triangle, the most stable form.

Consider The Following Poem:

I looked in the mirror, and saw my wife

I got married

My wife gave birth
I had a baby

Chapter Ten: The Gypsy Idiom

It is sometime in 1990 and I am heading up the road that winds through ancient caves of white stucco and red, ceramic tile floors, in the Gypsy Neighborhood of Granada, Spain called El Sacromonte (the sacred hill). I run into Manolin.

Manolin: "Que Hay?" (What is there?)

Me: "I am heading up to your cave to make something to eat." (It is where I was staying).

Manolin: "Saca la guitarra, y bajamos el monte." (Get out your guitar and we will go down the hill).

We go up to his cave and get my guitar, then head out of the Sacromonte, and descend by "The Paseo De Los Tristes" (The Pass of The Sad Ones) along the Rio Daro (the daro river), and head for "Plaza Nueva" "The New Plaza".

Along the way we stop at a café and bar.

Manolin: "Cuanto dinero tienes?" (How much money do you have?)

Me: "About 500 pesetas." (It was about five dollars).

Manolin: "Dame lo." (Give it to me.)

Me: If I do that, we will have no money for later.

Manolin: "Hombre, lo voy a doblar" (Man, I am going to double it).

He put the first coin in the slot machine and lost it. Then lost the second, the third, the fourth, the fifth.

Me: "Now that we have no money, what are we going to do?"

Manolin: "Buscamos la vida." (We will search for the life, or look for our means of existence in other words).

We arrive at Plaza Nueva and there is a man, American, sitting on a bench with his guitar by him in its case. Manolin stops in front of him and asks him if he will play something for us. He gets out his guitar and starts playing Ragtime Style.

Manolin: "Toca muy bien, no?" (He Plays very well, doesn't he?)

Me: "Si, muy, bella" (Yes, very beautiful)

The Guitarist: "How about if I buy you two a drink?"

We go into the nearest bar. The Guitarist orders a couple drinks for us. In Spain that comes with calamares (deep fried squid). We drink and eat, say our goodbyes, then Manolin and I head out, and through some old colonial, cobblestone street until we come across a bar where another Gypsy from the sacromonte was playing flamenco. We listen from outside until the Gypsy Guitarist ends the piece. He comes out of the bar and greets us:

Gypsy Guitarist: "Hola, primo, que hay?" (Hello, cousin, what is there?)

Manolin: "We came to see you play"

Gypsy Guitarist: "Primo, there is nothing I can teach you. Who is with you?" (Cousin, there is nothing I can teach you. Who is with you?)

Manolin: "A friend from America."

Gypsy Guitarist: "Claro, are you teaching him guitar?" (Of course, are you teaching him guitar?)

Manolin: "Claro." (Of course.)

We went our ways.

Me: "Es tu primo?" (He is your cousin).

Manolin: "Claro." (Of course)

We run into another guitarist.

The Other guitarist: Hola, Primo (Hello Cousin)

Manolin: "Hola, Primo."

The other guitarist: Is that your guitar?

Me: "yes"

The other guitarist: "Can I try it?"

I hand him the guitar.

The other guitarist: "I like it."

Me: "It is very old"

The Other Guitarist: "Claro." (Of course).

We again say our goodbyes and go our ways. Manolin and I head back up to the caves. We pass by an old woman on her terrace in front of her cave. She throws a rock at a dog, and yells at it: "Busca la vida por alli!" (Search for the life over there, or, search for your existence elsewhere, in other words.) The dog gives a yelp and runs off.

We come to a cave that is a bar owned by a Spanish woman. A young man, half Gypsy, and half Spanish, named Gabrielle, invites us in for a beer. We take our beers outside and sit on the step and watch the sun set behind the Alhambra, an old Moorish Castle across the valley. He asks me: "If Jesus was so great, then why did he die?"

I am later back in America, sitting on the porch in California with a young Bedouin man who recently found his way into my life. He has a leather bag, with a carry strap strung over his shoulder. He gets out a book inside, and says it is called "The Prophet", by Kahlil Gibran. He hands it to me. I open it and look through it.

Bedouin Man: "The student and the teacher are the same."

Me: "Of course."

Chapter Eleven: Introduction

In Gypsy Shamanism And The Universe, we told the best we could the experiences I had with a Gypsy Shaman that lead to astounding scientific discoveries, both in my works An Extraterrestrial Analysis and A Message From Extraterrestrials. Progress, it would seem, is at a standstill. I think it is because Gypsy Shamanism And The Universe is only half the story. It would seem when I was made into the AE-35 antenna and, after my second trip to Spain (Madrid) and my subsequent trips to Italy, it was omitted from my memory. However, I seem to have remedied that, and accessed it. Thus we now progress with the story of Gabriel.

Gabriel

Aside from the Gypsy Shaman, there was another Gentleman worth noting, with whom I actually spent more time. His name is Gabriel. He is actually half Gypsy and half Spanish. He lives in a cave among the other caves of the Sacromonte, including that of the Gypsy Shaman, Manuel.

Gabriel took the three of us under his wing in training for the guitar: The two children of the Gypsy Shaman, Manolin (was aged 27 at the time, which was around 1990) and his younger brother, Antonio (he must have been about 23).

Manolin and Antonio were at the cave of the Gypsy Shaman, Manuel, their father, where I was staying. Gabriel came by, and told Antonio to play bulerias. He said to, at a certain point, play g minor with the fifth removed (the fourth played instead) and began to sing. He sang with great power, "Tengo Un Caballo Blanco" which translates "I have a white horse". He asked me to hand him a Cordobes hat that was hanging on the wall. He put it on his head and said let's go. We, and all the other occupants, went together to a cave that was a bar. We all ordered drinks, and then went through one of the tunnels to a cave lit up green that had tables and posters of flamenco greats. He handed me the guitar and told me when he makes the chord on his arms, that I should play it on the guitar. It was a Deep Song G minor with the bottom e played in it. I did so when after he started singing and he played the chord on his arm at the point of crescendo.

We went back into the bar, it was getting dark, and there was a man playing guitar in the corner. Other people started to arrive at the bar, from other towns. We all went through another tunnel in the cave that lead to a chamber where there were wood chairs lining the walls. It was dark, lit only by candles. I sat all the way to the left side of the cave next to Antonio who was about to play bulerias for all the people in the chairs that lined the walls of the round cave. He played, along with others, and a person would get out his chair, stand in the center of the cave and sing, sit down after done and then someone else would then do the same, until one at a time the dancers would come to the center of the floor, dance, then sit down. I remember one of the set of lyrics that one singer sang:

Saca la red
Hermano, saca le red
Que ya salio la luna
No vaya coger

That is:

Take out the net
Brother, take out the net
The moon has already left
You are not going to catch it.

Gabriel tended to disappear for several days, and I did find out what that was about. He took me out of the Sacromonte one day just befor sunset, and we walked up to The Albaicin, the same Albaicin near where I bought the Gypsy Shaman his hose, the old Arabic City. We got to a plaza and entered a bar. Gabriel walked in, myself with him and approached the bar counter, looked briefly to his left, then shifted quickly to his right, then back to his left, then rapped out a fast flamenco rhythm on the counter. It was not long before someone ordered him a drink, which in Spain comes with tapas (appetizers) for free, usually Calamares (deep fried squid). Gabriel and I drank through the night and around sunrise we began to walk back to the Sacromonte, out of the old Arabic City and down through the cobblestone streets, leaning on one another the whole way down past people sweeping the sawdust out of their bar floors (bar floors are covered with saw dust to absorb the oil from parts of fish thrown to it: there are no trash cans, everything is thrown to the floor including napkins.) Gabriel offered to help one person with this task.

When we arrived at the Sacromonte, two women were approaching from the other direction. Gabriel said the one was his mother. I asked why he didn't say hi, and told me he did not want to embarrass her. He then went on to tell me that Tomatito (a famous Gypsy Flamenco Guitarist) was his half brother by way of his mother.

The next day I was talking to Antonio and he said that Gabriel was the brother of Tomatito. I said I know, by his mother. Antonio said no, by his father. I told him Gabriel told me by his mother. Antonio said Gabriel tiene historia. (He has a history)

The day after that I told Gabriel that Antonio says he is Tomatito's brother by his father and Gabriel said that it was due to being sort of traditional in an old fashioned way that Antonio said that.

This is interesting because one day when I was walking through the Sacromonte with Gabriel, we could see up in the hills from her cave patio, a Gypsy woman named Belen (Bethlehem) and I waved to her. Gabriel said to me that we had to have a talk. He took me up to a remote place in the hills and we sat on some rocks and he told me: "You don't know what you would be in for if you married a traditional Gypsy Girl".

It was the night before that I was up at her terrace with her father. He was not Manuel the Gypsy Shaman, but Manuel the singer, Deep Song singer (cantaor). I told Gabriel I was going there for a flamenco lesson and he said "Manuel es bueno" (Manuel is good). That is what everyone always said about him. It is best explained by Bob Dylan's lyrics, "His clothes are dirty but his hands are clean."

Chapter Twelve: The Sequence

We considered the ratio nine to five, then the proportion and found it in Saturn Orbit to Jupiter orbit, Solar Radius to Lunar Orbit, Gold to Silver and if flower petal arrangements. It is left then to consider the whole number multiples of nine-fifths (1.8) or the sequence:

1.8, 3.6, 5.4, 7.2,...

in other words, and we look to see if it is in the solar system and find it is in the following ways:

1.8

Saturn Orbit/Jupiter Orbit
Solar Radius/Lunar Orbit
Gold/Silver

3.6

(10)Mercury Radius/Earth Radius
(10)Mercury Orbit/Earth Orbit

(earth radius)/(moon radius)=
4(degrees in a circle)(moon distance)/(sun distance)
= 3.7 ~ 3.6

There are about as many days in a year as degrees in a circle.

(Volume of Saturn/Volume Of Jupiter)(Volume Of Mars) = 0.37 cubic earth radii
~ 3.6

The latter can be converted to 3.6 by multiplying it by (Earth Mass/Mars Mass) because Earth is about ten times as massive as Mars.

5.4

Jupiter Orbit/Earth Orbit
Saturn Mass/Neptune Mass

7.2

10(Venus Orbit/Earth Orbit)

Chapter 13: The Neptune Equation

If we consider as well the sequence where we begin with five and add nine to each successive term: 5, 14, 23, 32…Then, the structure of the solar system and dynamic elements of the Universe and Nature in general are tied up in the two sequences:

5, 14, 23, 32,…

and

1.8, 3.6, 5.4, 7.2,…

How do we find the connection between the two to localize the pivotal point of the solar system? We take their difference, subtracting respective terms in the second sequence from those in the first sequence to obtain the new sequence:

3.2, 10.4, 17.6, 24.8,…

Which is an arithmetic sequence with common difference of 7.2 meaning it is written

$7.2n - 4 = a_n$

The a_n is the nth term of the sequence, n is the number of the term in the sequence.

This we notice can be written:

[(Venus-orbit)/(Earth-orbit)][(Earth-mass)/(Mars-mass)]n – (Mars orbital #) = a_n

We have an equation for a sequence that shows the Earth straddled between Venus and Mars. Venus is a failed Earth. Mars promises to be New Earth.

The Mars orbital number is 4. If we want to know what planet in the solar system holds the key to the success of Earth, or to the success of humans, we let n =3 since the Earth is the third planet out from the Sun, in the equation and the result is a_n = 17.6. This means the planet that holds the key is Neptune. It has a mass of 17.23 earth masses, a number very close to our 17.6.

Not only is Neptune the indicated planet, we find it has nearly the same surface gravity as earth and nearly the same inclination to its orbit as earth. Though it is much more massive than earth, it is much larger and therefore less dense. That was why it comes out to have the same surface gravity.

Chapter 14: The Uranus Equation

I asked what needs to be done to solve My Neptune Equation, by going deep with the guitar in Solea Por Buleras. I found the answer was that I didn't have enough information to solve it.

Then I realized I could create the complement of the Neptune equation by looking at the Yang of 5/3, since the Neptune equation came from the Yin of 9/5.

We use the same method as for the Neptune equation:

Start with 8 and add 5 to each additional term (we throw a twist by not starting with 5)

5/3 => 8, 13, 18, 23,...

List the numbers that are whole number multiples of 5/3:

5/3n = 1.7, 3.3, 5, 6.7,...

Subtract respective terms in the second sequence from those in the first:

6.3, 9.7, 13, 16.3,...

This is an arithmetic sequence with common difference 3.3. It can be written:

$(a\_n) = 3 + 3.3n$

This can be wrtten:

Earth Orbital # + (Jupiter Mass/Saturn Mass)n = a_n

Letting n = 3 we find $a\_n = 13$

The closest to this is the mass of Uranus, which is 14.54 earth masses. If Neptune is the Yin planet, then Uranus is the Yang planet. This is interesting because I had found that Uranus and Neptune were different manifestations of the same thing. I had written:

I calculate that though Neptune is more massive than Uranus, its volume is less such that their products are close to equivalent. In math:

N_v = volume of Neptune
N_m = mass of Neptune
U_v = volume of Uranus
U_m = mass of Uranus

$$(N\_v)(N\_m) = (U\_v)(U\_m)$$

Chapter 15: The Earth Equation

We then sought the Yang of six-fold symmetry because it is typical to physical nature, like snowflakes. We said it was 5/3 since it represents the 120 degree measure of angles in a regular hexagon and we built our universe from there, resulting in the Uranus Integral, which was quite fruitful. Let us, however, think of Yang not as 5/3, but look at the angles between radii of a regular hexagon. We have:

$360 - 60 = 300$

$300 + 360 = 660$

$660/360 = 11/6$

We say Yin is 9/5 and Yang is 11/6 and stick with The Gypsy Shaman's 15 (See An Extraterrestrial Analysis, chapter titled "Gypsy Shamanism And The Universe") and build our Cosmology from there.

We already built The Neptune Equation from 9/5 and used it with 5/3 to derive the planet Europia, but let us apply 11/6 in place of 5/3:

$11/6 => 11/6, 11/3, 11/2, 22/3,\ldots = 1.833, 3.667, 5.5, 7.333,\ldots$

$11/6 => 6, 6+11 = 17, 17+11=28, 28+11=39, \ldots = 6, 17, 28, 39,\ldots$
Subtract the second sequence from the first:
$4.167, 13.333, 22.5, 31.667,\ldots$
Now we find the common difference between terms in the latter: $9.166, 9.167, 9.167,\ldots$

$(a\_n) = a + (n-1)d = 4.167+(n-1)9.167 = 4.167 + 9.167n - 9.167 = 9.167n-5$

Try n=3: $9.167(3) - 5 = 27.501 - 5 = 22.501$ (works)
Our equation is:

$(a\_n) = 9.167n - 5$

We notice this can be written:

$[(Saturn\ Orbit)/(Earth\ Orbit)]n - (Jupiter\ Orbital\ \#) = (a\_n)$

The Neptune Equation for n=3 gave Neptune masses, the Uranus equation for n=3 gave Uranus masses. This equation for n=3 gives close to the tilt of the Earth (23.5 degrees) in a form that is exactly half of the 45 degrees in a square with its diagonal drawn in. In the spirit of our first cosmology built upon 9/5, 5/3, and 15, we will call this equation The Earth Equation.

Chapter: 16: The Unification Of Pi and Phi by Nine-Fifths

All that is left to do is to consider pi, the circumference of a circle to its diameter, and phi, the golden ratio, since they are two most important, if not most beautiful ratios in mathematics.

I have found nine-fifths occurs throughout nature in the rotation of petals around a a flower for a most popular arrangement, in the orbits of jupiter to saturn in their closest approaches to the sun, in the ratio of the molar masses of gold to silver, and in the ratio of the solar radius to the lunar orbit. I now further go on to say that this nine-fifths unifies the two most important ratios in mathematics pi and the golden ratio (phi), in that

pi + phi = 3.141 + 1.618 = 4.759

Because the numbers after the decimal in the sum (the important part) are 5 and 9 and 7, the average of 5 and nine.

I should also like to point out that the fourth and fifth numbers after the decimal in pi are 5 and 9 and in phi the second and third numbers after the decimal are one and eight where nine-fifths divided out is one point eight, and, further, the first and second numbers after the decimal in phi add up to make 7, the average of 9 and 5, and subtract to make five, and the second and third digits after the decimal add up to nine. So not only does the solar system unify pi and phi through nine-fifths, pi and phi taken alone express nine-fifths in the best possible ways.

Chapter 17: Summary About Nine-Fifths

In Summary, this most beautiful constant in Nature, of nine-fifths, which unifies pi and the golden ratio, the two most beautiful ratios in mathematics, is the most popular arrangement of petals around a flower, and is in the ratio of the molar masses of gold to silver, the most precious of the metals. One could only guess that such would be noticeable throughout the universe and would be noticed by other intelligent life in the universe, and, as such the relationship would be ideal for transmission of a message by other intelligent life forms in the universe to say to the receiver: "I am here and have noticed it". The curious thing is that it exists in our solar system in the ratio of the orbit of Saturn at its closest approach to the sun compared to the orbit of Jupiter at its closest approach to the sun, mystically putting the Earth at one unit from the Sun (Jupiter and Saturn being the largest and most massive in the solar system, the Earth that planet which harbors intelligent life in the Solar System), and in the ratio of the Sun's radius to the orbital radius of the moon (moon-earth separation). The question is why does this relationship for pi and the golden ratio of nine-fifths not just exist throughout the Universe in gold, silver, and perhaps flower petal arrangements, but more locally in our solar system? Does it mean humans have some kind of divine specialness in the cosmos? While we can know that the nine-fifths phenomenon would be represented by gold and silver throughout the universe, we currently cannot know whether the inhabitants of other star systems have it in their sun (star) they orbit and a moon that might orbit their planet or in the ratio of the orbits of the largest planets in their star system putting their planet at one unit from their star. The answer to this enigma, I think, must be of extraordinary profundity.

Chapter 18: (pi) and (e)

I have talked about how 9/5, which I have found exists in Nature and the Universe, unifies pi and the golden ratio (phi):

(pi) + (phi) = 3.141 + 1.618 = 4.759

because the first three numbers after the decimal are 7, 5 and 9. Seven is the average of nine and five, and the second number is our 5 in nine-fifths and the third number is the 9 in nine-fifths.

It would seem 9/5 unifies euler's number, e, and pi, as well:

(pi) + (e) = 3.141 + 2.718 = 5.859

The second number after the decimal is the 5 in nine-fifths, and the third number after the decimal is the 9 in nine-fifths. The first number after the decimal is eight. This is significant because the 8 is the 8 in 1.8, which is 9/5 divided out. The one will take you all the way around a circle, what is left is 0.8.

Ian Beardsley
January 17, 2013

Chapter 19: How Telepathy Works

Much in the same way that the Gypsy Shaman, or Chovihano, gave me a deep understanding of the number 15 by having me buy him a new hose I never lost, but that he took himself, he gave me another gift, the gift of telepathy.

When I helped him install an antenna for television reception, since I had become him and he had become me when I was with him setting down the hose I bought him in a chamber in his cave, I could know what he thought of my thoughts, and, he could know what I thought of his. In this sense telepathy was not a real, physical telepathy, but one the result of psychology, but it was equivalent to the real thing.

Ian Beardsley
February 5, 2014

Data For The Planets

planet	Orbit (O)	Radius (R)	Mass (M)
mercury	0.387099	0.382	0.0558
venus	0.723332	0.949	0.8150
earth	1.000000	1.000	1.0000
mars	1.523691	0.532	0.1074
jupiter	5.202803	11.27	317.893
saturn	9.53884	9.44	95.147
uranus	19.1819	4.10	14.54
neptune	30.0578	3.88	17.23

O for Earth = 1.495979E13 cm   R for Earth = 6,378 km   M for Earth = 5.976E27 g

Earth-Moon Separation: 3.84E10 cm
Solar Radius: 6.9599E10 cm

Molar Mass of Gold: Au = 196.97
Molar Mass of Silver: Ag = 107.87

Saturn (minimum distance from sun) = 9.014 AU = 1.348E9 km
Jupiter (minimum distance from sun) = 4.951 AU = 7.409E8 km

Jupiter (maximum distance from the sun): 5.455 AU ~ 5.4 Astronomical Units

69

The Yang Elements

K = potassium
Na = Sodium

K/Na = 39.10/22.99 = 1.71

5/3 =1.67 ~ 1.7

K/Na ~ 5/3

Chapter 20: Discover

Back in 2005, as I did my research, I developed a different convention for rounding numbers than we use. I felt I only wanted to use the first two digits after the decimal in processing data using molar masses of the elements. This I did, unless a fourth digit less than five followed the third digit in my calculations, then, I would use the first three digits for greater accuracy. Now I am taking the introductory class in computer science at Harvard, online, CS50x. Working in binary, where all numbers are base two, I see that it was no wonder I got the results I did, on the first try when I wondered if the golden ratio conjugate, 0.618 to three places after the decimal would be in artificial intelligence (AI) since it is recurrent throughout life.

I was taking polarimetric data on the eclipsing binary Epsilon Aurigae at Pine Mountain Observatory in the 1980's, for which there was a paper in the Astrophysical Journal upon which my name appears as coauthor, while studying physics at The University of Oregon. As well I was studying Spanish, and in an independent study project through the Spanish Department, I left the University to live of among the caves of the Gypsies of Granada, Spain. In doing as such, I disappeared from the entire world, only to return from another kind of life finding the world was now a much different place. Around 2005, I enrolled in chemistry at Citrus College in Southern California, when I did the following:

If the golden ratio conjugate is to be found in Artificial Intelligence, it should be in silicon, phosphorus, and boron, since doping silicon with phosphorus and boron makes transistors.

We take the geometric mean between phosphorus (P) and Boron (B), then divide by silicon (Si):

$$\sqrt{PB}\,/\,Si = \sqrt{(30.97)(10.81)}\,/\,28.09 = 0.65$$

Harmonic mean between phosphorus and boron divided by silicon:

$$\frac{2PB}{P+B}\,/\,Si = \frac{2(30.97)(10.81)}{30.97+10.81}\,/\,28.09 = 0.57$$

Arithmetic mean of these two numbers: (0.65 + 0.57)/2 = 0.61

0.61 is the first two digits of the golden ratio conjugate.

Now the golden ratio conjugate is in the ratio of a persons height and the length from foot too navel, and is in all of ratios between joints in the fingers, not to mention that it serves in closest packing in the arrangement of leaves around a stem to provide maximum exposure to sun and water for the plant. Here we see that the golden ratio is not in artificial intelligence which is 0.62 to two places after the decimal, but that the numbers in its value are in artificial intelligence 0.61, which is 0.618 to three places after the decimal. That is, if we consider the first two digits in the ratio. If we consider the golden ratio conjugate to one place after the decimal, which is 0.6, then we say artificial intelligence does have the golden ratio in its transistors. I like to think of I, Robot by Isaac Asimov, where in one of that collection of his short stories, robots are not content with what they are, and need more: an explanation of their origins. They can't believe that they are from humans, since they insist humans are inferior. Or, I like to think of the ship computer HAL in 2001, he mimics intelligence, but we don't know if he is really alive. Perhaps that is why to two place after the decimal, AI carries the digits, but is not the value.

In any case, I have written a program called Discover that would enable one to process arithmetic, harmonic, and geometric means for elements or whatever, because someone, including myself, might want to see if there are any more nuances hidden out there in nature. I have already found something that seems to indicate extraterrestrials left their thumbprint in our physics. I even find indication for the origin of a message that would seem they embedded in our physics. That origin comes out to be the same place as the source of the SETI Wow! Signal, Sagittarius. The Wow! Signal was found in the Search For Extraterrestrial Intelligence and a possible transmission from ETs. But that is another subject that is treated in my book: All That Can Be Said at, http://issuu.com/eanbardsley/docs/allsaidonline

I now leave you with my program, Discover in the language C, with a sample running of it:

The Program Discover

```c
#include <stdio.h>
#include <math.h>
int main(void)
{
printf("transistors are Silicon doped with Phosphorus and Boron\n");
printf("Artificial Intelligence would be based on this\n");
printf("the golden ratio conjugate is basic to life\n");
printf("The Golden Ratio Conjugate Is: 0.618\n");
printf("Molar Mass Of Phosphorus (P) Is: 30.97\n");
printf("Molar Mass Of Boron (B) Is: 10.81\n");
printf("Molar Mass Of Silicon (Si) Is: 28.09\n");
int n;
do
{
printf("How many numbers do you want averaged? ");
scanf("%d", &n);
}
while (n<=0);

float num[n], sum=0.0, average;
for (int i=1; i<=n; i++)
{
printf("%d enter a number: ", i);
scanf("%f", &num[n]);
sum+=num[n];
average=sum/n;
}
printf("sum of your numbers are: %.2f\n", sum);
printf("average of your numbers is: %.2f\n", average);

float a, b, product, harmonic;
printf("enter two numbers (hint choose P and B): \n");
printf("give me a: ");
scanf("%f", &a);
printf("give me b: ");
scanf("%f", &b);
product = 2*a*b;
sum=a+b;
harmonic=product/sum;
printf("harmonic mean: %.2f\n", harmonic);
```

```c
double geometric;
geometric=sqrt(a*b);
printf("geometic mean: %.2f\n", geometric);

printf("geometric mean between P and B divided by Si: %.2f\n",
geometric/28.09);
printf("harmonic mean between P and B divided by Si: %.2f\n", harmonic/28.09);

printf("0.65 + 0.57 divided by 2 is: 0.61\n");
printf("those are the the first two digits in the golden ratio conjugate\n");
}
```

Running Discover

```
jharvard@appliance (~): cd Dropbox/pset2
jharvard@appliance (~/Dropbox/pset2): ./add
transistors are Silicon doped with Phosphorus and Boron
Artificial Intelligence would be based on this
the golden ratio conjugate is basic to life
The Golden Ratio Conjugate Is: 0.618
Molar Mass Of Phosphorus (P) Is: 30.97
Molar Mass Of Boron (B) Is: 10.81
Molar Mass Of Silicon (Si) Is: 28.09
How many numbers do you want averaged? 2
1 enter a number: 9
2 enter a number: 5
sum of your numbers are: 14.00
average of your numbers is: 7.00
enter two numbers (hint choose P and B):
give me a: 30.97
give me b: 10.81
harmonic mean: 16.03
geometic mean: 18.30
geometric mean between P and B divided by Si: 0.65
harmonic mean between P and B divided by Si: 0.57
0.65 + 0.57 divided by 2 is: 0.61
those are the the first two digits in the golden ratio conjugate
jharvard@appliance (~/Dropbox/pset2):
```

Chapter 21

Once you realize nine-fifths is not just at the crux of Gold and Silver, Pi and the Golden Ratio, Pi and Euler's Number e, the five-fold symmetry that is typical of life, Jupiter and Saturn, Sun and Moon, it is not long before you realize its compliment is 5/3 and that you form the sequences:

(For 9/5)   5, 14, 23, 32,... and 1.8, 3.6, 5.4, 7.2,...
(For 5/3)   8, 13, 18, 23,... and 1.7, 3.3, 5, 6.7,...

For which you get:

$7.2n - 4 = a\_n$ and $(a\_n) = 3 + 3.3n$ respectively.

In the latter, letting the 3.3 be Earth Gravities rounded to the nearest ten (980), we have:
(v) = 2,940 cm/s + (3,234 cm/s/s)t
This is the differential equation:

(dx) = (2,940 cm/s)dt + (3,234 cm/s/s)t dt

$$\int_0^{15}(2{,}940cm/s)dt + \int_0^{15}(3{,}234cm/s/s)tdt = 4.07925km \quad \text{(Manuel's Integral)}$$

15 seconds because there are 15 degrees in an hour of right ascension. The factor of one fourth enters because the kilometer is defined by the distance from the pole to the equator, not by the circumference of the Earth. Notice the 0.07925 has the nine and five of nine-fifths, the average of nine-fifths and the 2 used to make it.
Mach 1 = 768 mph =1,235 km/hour
That is mach 1 in dry air at 20 degrees C (68 degrees F) at sea level.
If we write, where 1,235 km/hr (mach 1) = 0.343 km/s, then:
34,300 cm/s =2,940 cm/s + (3234 cm/s/s)t
and
t=9.696969697 seconds = 9 23/33 s = 320/33 seconds ~ 9.7 seconds
So, the Uranus equation is a time of 9.7 seconds to reach mach 1. Putting that time in the integral:
(x) = (2,940)(320/33) + 1/2(3234(320/33)^2 = 180,557 cm 1.80557 km ~ 1.8km
Thus we see Manuel's Integral reaches mach one in about 9.7 seconds after traveling a distance of about 1.8 kilometers. Let's convert that to miles:
(1.8km)(0.621371mi/km)=1.118478 miles
Manuel's Integral reaches mach one in 1.8 kilometers, which is the amount of kilometers in a mile and is the 9/5 that occurs in Nature and the Universe, not to mention that it unifies pi and golden ratio and pi and Euler's number e. It is one compact statement that embodies everything and connects it to Earth Gravity. – Ian Beardsley, August 15, 2013

Manuel's Second Integral

Earth gravity (g) is 9.81 m/s/s
This is close to 9.80 m/s/s
Indeed if rounded it to one place after the decimal, it would be 9.8 m/s/s
This value when converting to cm/s/s gives g = 980 cm/s/s
There may be good reason to write it like this (which is rounding it to the nearest ten) because we see in our research that it is fruitful not mention that it provides a nice form for the value if we want to create a new system of units both with a zero at the end for the value and that is connected to nature, which it is, in Manuel's integral. Also the nine is the nine in the nine-fifths connected to nature and mathematical constants, as we have shown in our research, and the eight is the 0.8 in the 1.8 that is nine fifths, the fraction around a circumference of a circle that is nine-fifths of a circumference.

Let us consider the Neptune Equation:

7.2x −4 = y

Let 7.2 be Earth Gravities:

(v) = 7.2t −4
(dx/dt) = 7.2t −4
(dx) = 7.2t dt − 4dt
(7.2)(980 cm/s/s) = 7,056 cm/s/s

v=4=7.2t
t=(5/9)
(7,056)(5/9)=v_0 = 3,920 so,

(dx) = 7,056 cm/s/s t dt −3,920 cm/s dt

$$\int_0^{15} 7{,}056 cm/s/s\, t\, dt - \int_0^{15} 3{,}920 cm/s\, dt = 7.35 km = \frac{147}{20} km$$

We call this Manuel's Second Integral

Ian Beardsley
November 26, 2013

Manuel's Third Integral

Yin of 9/5 (five-fold symmetry)

360/5 = 72
360-72=288
288+360=648
648/360=9/5

Yang of 5/3 (six-fold symmetry)

360/6=60
360-60=300
300-60=240
240/360=2/3
2/3+1=5/3

Yang 2 of 11/6 (six-fold symmetry)

360-60=300
300+360=660
660/360=11/6

We consider the earth equation:

(y) = 9x-5
v=9t-5
g=980 cm/s/s
9(980)=8,820 cm/s/s

5=9t
(t)=5/9

(v_0) = (8,820 cm/s/s)(5/9) =4,900 cm/s

dx = (8,820 cm/s/s)t dt – (4,900 cm/s) dt

$$\int_0^{15} 8,820cm/s/stdt - \int_0^{15} 4,900cm/sdt = 9.1875km$$

The Integral has the 9 of nine-fifths, the five, the seven, which is the average of 9 and 5 and the one and eight of the 1.8 that is 9/5 divided out. We call this Manuel's Third Integral.

We have the Neptune Equation:

7.2x –4

We have the Uranus Equation

3.3x + 3

And now with our alternate cosmology we have The Earth Equation:

9x-5

With three equations we can write the parameterized equations in 3-dimensional space, parameterized in terms of t, for x, y, and z. We can write from that f(x,y,z) and find the gradient vector, or normal to the equation of a plane in other words, and from that a region in space.

$$x(t) = \frac{36}{5}t - 4$$

$$y(t) = \frac{33}{10}t + 3$$

$$z(t) = 9t - 5$$

$$\frac{5x + 20}{36} = \frac{10y - 30}{33} = \frac{z + 5}{9}$$

$$\frac{5}{36}x - \frac{10}{33}y - \frac{1}{9}z + \frac{10}{11} = 0$$

$$\nabla f = \langle 5/36, -10/33, -1/9 \rangle$$

a=5/36   b=-10/33

$$c = \sqrt{(5/36)^2 + (10/33)^2} = \sqrt{0.0918 + 0.019} = 0.3328$$

d=-1/9

$$\tan \alpha = b/a$$

$$\alpha = -65.358°$$

$$\tan \beta = d/c$$

$$\beta = -18.46°$$

-65.358 degrees/15 degrees/hour =-4.3572 hours

24 00 00 – 4.3572 = 19.6428 hours

RA: 19h 38m 34s
Dec: -18 degrees 27 minutes 36 seconds

$$\left\langle \frac{5}{4}, -\frac{30}{11}, -1 \right\rangle \cong \left\langle 1, -3, -1 \right\rangle$$

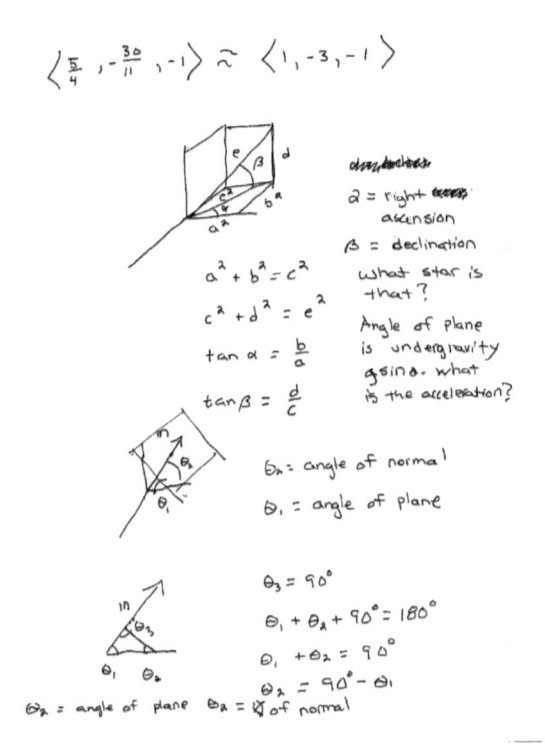

a = right ascension

β = declination

what star is that?

Angle of plane is under gravity $g\sin\theta$. what is the acceleration?

$a^2 + b^2 = c^2$

$c^2 + d^2 = e^2$

$\tan\alpha = \frac{b}{a}$

$\tan\beta = \frac{d}{c}$

θ_2 = angle of normal

θ_1 = angle of plane

$\theta_3 = 90°$

$\theta_1 + \theta_2 + 90° = 180°$

$\theta_1 + \theta_2 = 90°$

$\theta_2 = 90° - \theta_1$

θ_2 = angle of plane θ_2 = ∠ of normal

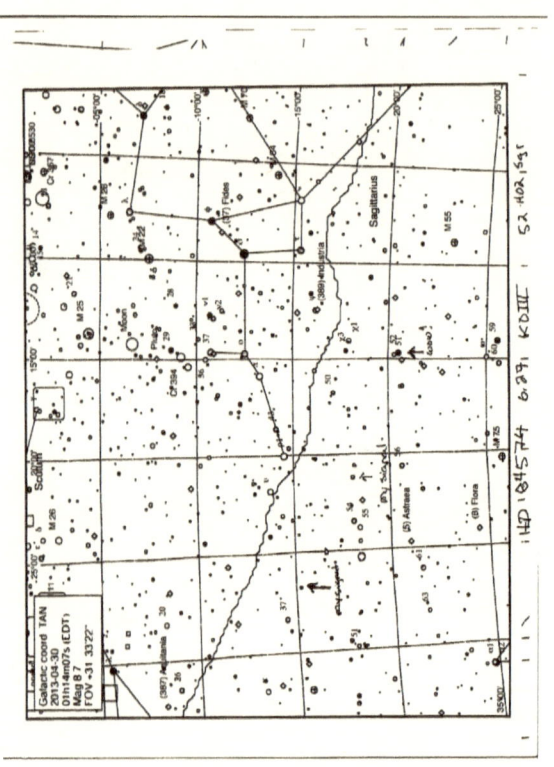

The projection by my calculation through my cosmology of yin, yang, and 15 for the origin of my message from extraterrestrials was somewhere in the easternmost part of the constellation Sagittarius. This happens to be the same place where the one possible alien signal was detected in the Search For Extraterrestrial Intelligence (SETI). It was called "The Wow Signal" because on August 15, 1977 the big ear antenna received something that seemed not like star noise, but exactly what they were looking for in an extraterrestrial signal. Its name is what it is because the astronomer on duty, Jerry R. Ehman wrote "Wow!" next to the numbers when they came in. Incredibly, it lasted the full 72 seconds that the Big Ear antenna listened for it. I say incredible because I have mentioned the importance of 72, not just in my Neptune equation – for which my location in space was derived in part – but because of its connection to the Gypsy Shaman's AE-35 antenna and its relation to 72 in the movie "2001: A Space Odyssey". The estimation of the coordinates for the origin of the Wow signal are two:

19h22m24.64s

19h25m17.01s

With declination of:

-26 Degrees 25 minutes 17.01 s

That is about 2.5 degrees from the star group Chi Sagittarri

It is very close to my calculation for an extraterrestrial civilization that I feel hid a message in our physics, which I calculate to be near HD 184835 and exactly at:

19h 38m 34s
-18 Degrees 27 minutes 36 seconds

The telescope that detected the Wow Signal was at Ohio Wesleyan University Delaware, Ohio called The Perkins Observatory.

Ian Beardsley May 6, 2013

A diode is made of silicon (sand) doped with phosphorous and boron, all naturally occurring elements made (forged) in interior of stars. A silicon diode needs 0.6 volts to conduct a current. In a diode gate a high voltage is six volts and the circuit is on and a low voltage, or ground (zero), the circuit is off. Off is coding for zero and on is coding for 1. Zeros and ones can be strung together to make numbers, and letters in the alphabet.

Manuel, the Gypsy Shaman, said: "God is an Idea, There are two elements." We can take this to be our Yin and Yang of 9/5 and 5/3. Low voltage and the circuit is off; high voltage and the circuit is on. These, again, are the two elements of the Shaman's yin and yang.

Because you have to run a diode at least 0.6 volts, this is subtracted from the six volts that turn the diode on, so really it operates on slightly less than six volts. This voltage drop of 0.6 Volts in the circuit is what we want to look at. We will evaluate the voltage drop over the Shaman's mysterious 15 seconds and obtain an amount of power, then apply that power over one year – the time it takes the earth to go around the Sun. We have already found that the Shaman's 15 not only unifies meters with the universe, but meters with feet, amazingly in the Uranus Integral outlined both in The Exploits Of Manuel and Further Exploits Of Manuel.

Now we do the calculation:

One electron volt is the energy of an electron falling through a potential difference of one volt, which is equal to 1.602E-19 Joules.

One Joule is 10,000,000 ergs

(0.6 Volts)(1.602E-19 Joules) = 9.612E-20 Joules

(9.612E-20 Joules)(10,000,000 ergs/Joule) = 9.612E-13 ergs

(9.612E-13ergs)/(15 seconds) = 6.408E-14 ergs/second

365.25 * 24 * 60 * 60 = 31,557,600 seconds/year

(6.408E-14 ergs/sec)(31,557,600 seconds) = 2.022211008 microergs = 2.0 microergs

It is almost exactly two microergs! (Is the two microergs prompting that we decode the message in binary?)

What Message?

My book A Thorough Dimensional Analysis is about how I provide evidence in support of the idea that extraterrestrials gave us our units of measurement, like the foot, the meter, the second, in Ancient times despite the history that tells us the metric system is modern, and the foot-pound system, perhaps a little ambiguous in its history. I think along the way in doing this I may have found an encrypted message from extraterrestrials in the number:

2.022211008

Insofar as it came to us from natural constants like the charge of an electron, the orbital period of the Earth, and numbers we invented like the measure of electromotive potential, the volt. The idea is that extraterrestrials somehow told us to measure electromotive potential with a unit of volts because it would produce the above number. I am interested in it because of the repetition of ones and twos, and the separation of numbers by zeros.

Let us look at the part after the decimal. It is:

022211008

It is the second number three times, the first number two times and the last number is the number of digits that precede it.

The three twos and the two ones add up to the last digit, eight.

The two ones add up to each of the three twos.

It counts from right to left 0,1,2,...

Decoding The Message

The message is:

022211008

The 222 adds up to six, which in binary is 110. The next segment is 1100, which is a nibble, and is the number 12. This segment is in binary to begin with. The last digit is an eight, which is 1000 in binary. Thus we have, including the first digit, a zero, which is a bit:

(0) (110) (1100) (1000)

We write this as three nibbles:

(0110) (1100) (1000)

Thus 022211008 translates as 0110 1100 1000

Which are the three numbers:

6, 12, 8

These three numbers divide nicely into the 360 degrees of a circle:

360/12 = 30 360/6 = 60 360/8 = 45

These are the angles in the three special triangles with which we can write out the trigonometric functions of important angles in closed form:

30-60-90 60-60-60 45-45-90

6, 12, 8 have the common denominator of two. This gives us the three numbers 3, 6, 4:

364

Which is very close to the number of Earth days in a year (365). Are extraterrestrials pointing out that the number of Earth days in a year are closely connected to the three special triangles? The number in binary is interesting, because the ones scroll off from right to left nibble to nibble, one digit at a time. The message is also that the number of days in a year are connected to the three regular tessellators. They have sides 3, 6, and 4.

Is The Extraterrestrial Message Real?

I was lead in my research to think there was good reason we should try to create a system of units for measurement of physical reality that not only related the foot-pound system to the metric system but that connected to Nature in a profound way, or, if not, use the connection between the two systems to aid us in finding one that is more profoundly connected to nature than anything we have so far.

Interestingly, I had suggested we should not change the unit of time, one second, because I believed it was already connected to nature in a profound way, as I outlined in my book SETI: Another Signal In Sagittarius in the chapter at the beginning titled The Dance Of The Heavens.

I find that interesting because that profound connection lay in the connection of the three special triangles to the 365 day year in that 365 is close to the 360 degrees in a circle, which means each day of the year represent about one degree of motion of the Earth around the Sun.

I now realize that as well, the importance of my hypothetical message from extraterrestrials is in the same idea: That, it connects the Earth year to the three special triangles. Does this support the idea that the hypothetical message from extraterrestrials is real and that we should really look for a system of measurement that is more profoundly connected with nature?

Ian Beardsley
December 3, 2013

The Pluto Equation

We will call it The Pluto Equation since n=3 (earth is the third planet from the sun) yields 38.9, 38.9 being closest to the distance of Pluto from the Sun in astronomical units which is 39.44 astronomical units.

Yin = $\frac{9}{5}$ Yang = $\frac{5}{3}$

alteranate cosmology

Yin = $\frac{9}{5}$ Yang = $\frac{11}{6}$

Alternate ~~Yangy~~ Yin

Angle in a regular pentagon is

$108°$

$360° - 108° = 252°$

$360° + 252° = 612°$

$\frac{612°}{360°} = \frac{17}{10} = 1\frac{7}{10} = 1.7$

$\frac{17}{10} n$

$1.7, 3.4, 5.1, 6.8, \ldots$

$1.7 \quad 1.7$

$10, 10+17 = 27, 27+17 = 44, 44+17 = 61, \ldots$

$\Rightarrow 10, 27, 44, 61, \ldots$

$10-1.7 = 8.3, 27-3.4 = 23.6, 44-5.1 = 38.9,$

$61 - 6.8 = 54.2$

$\Rightarrow 8.3, 23.6, 38.9, 54.2$

$15.3 \quad 15.3 \quad 15.3$

Common difference $= 15.3$

$a_n = a+(n-1)d \quad a = 8.3, \quad d = 15.3$

$a_n = 8.3 + (n-1)15.3 = 8.3 + 15.3n - 15.3$

$\boxed{a_n = 15.3n - 7} \quad n=3 \Rightarrow 38.9$

$$\sqrt{a^2 + b^2 + d^2} = \text{distance to signal}$$
(in light years?)

Pluto equation being fourth equation is time parameterized in terms of T and gives the time for the wow signal (years after SETI founding?)

$$a = \frac{5}{36}, \quad b = -\frac{10}{33}, \quad d = -\frac{1}{9}$$

$$\sqrt{0.019 + 0.0918 + 0.0123} = 0.1231 \text{ light years}$$

Pluto equation;

$$T = 15.3 t - 7$$

$$\frac{T + 7}{15.3} = t \qquad \frac{T}{15.3} = \text{time element}$$

because our vector is:

$$\nabla F = \left\langle \frac{5}{36}, -\frac{10}{33}, -\frac{1}{9}, -\frac{10}{153} \right\rangle \qquad \frac{\frac{1}{15.3}}{10} = \frac{10}{153}$$

A Time For Contact

First we found a message embedded in our physics, then we used our system of three equations to find its origin in space. It turned out to be from the same place in space as the SETI Wow! Signal, Sagittarius.

We do have a fourth equation, which represents time. With this we try to find when the extraterrestrials might try to make contact with us again.

We received the SETI signal on August 15, 1977. I discovered my signal around May 5, 2013. The fourth equation from which we derive a time, I call The Pluto Equation, derived in my book SETI: Another Signal In Sagittarius. It is:

$T = 15.3t - 7$

This gives the time element in our vector of -1/15.3

May 5, 2013 – August 15, 1977 = 36 years 3 months 7.2 days

Is the time I estimate between the SETI Signal and mine rounding months to 30 days and so forth. This is 36.27 years by my reckoning. We reduce this by our factor of 15.3 (which we note the 15 is the degrees the earth rotates through in an hour and the 3 is the earth planet number). 36.27/15.3=2.37 years.

2013 + 5 months (may) + 2.37 years=2013+0.41 years+2.37 years
=2015.78 years

(0.78)(12)=9.36 months and the theorized date of contact is around September, 2015. This is an estimate, and it may be there is another way of doing this.- Ian Beardsley, Dec 18, 2013

Another Projection For Extraterrestrial Contact

There is another way of calculating when the possible message from extraterrestrials in Sagittarius, the SETI Wow! Signal, will repeat it self. Our discovery of another message from the same place began with the Gypsy Shaman's hose collection of 15 hoses making us realize that 15 was important because the Earth rotates through 15 degrees in an hour, and 15 seconds lead to the dynamic integral, Manuel's Integral. The first message was on August 15, 1977. We noted that that 15 of August pointed to the Shaman's 15 hoses and the two sevens in 1977 add up to 14, which when added to the one in 19 is 15 as well, while the 9 is the nine of nine-fifths that we found in Nature from which we calculated a place in Sagittarius where the SETI Wow! Signal is. We decoded another message on around May 5 of 2013. The next message should then be on August 15, 2015, to line up the Shaman's fifteens. August is a good time to view the constellation Sagittarius from where I am in Southern California. Sagittarius has always been my favorite constellation, because it is in the center of the Galaxy, rich with globular, and open clusters that can be viewed with binoculars. Summer is when stargazing becomes exciting because you get both a rich sky and warm, uplifting weather. Also, August is the eighth month and our nine-fifths divided out is one point eight. The one takes you around a circle once, leaving point eight.

Dec 24, 2013

Chapter 22: A Thorough Treatment Of Dimensional Analysis

Units are the basic magnitudes we use to measure dimensions, which are quantities we use to describe nature, like length, mass, time, temperature, and energy. We define a kilometer to one ten thousandth of the distance from the pole to the equator, a meter to be one thousandth of a kilometer, a centimeter to be one hundredth of a meter. We define a second to be one 86, 400th of an Earth day, and a gram to be the weight of a cubic centimeter of water at standard temperature and pressure, which is zero degrees centigrade and the pressure at sea level. We define one degree centigrade by saying zero degrees is the temperature at which water freezes, and 100 degrees centigrade is the temperature at which water boils. This is the metric system, and we created it in terms of the world we know. And we created it so that by assigning a Greek prefix, like Kilo (meaning thousand) when attached to meter, means a thousand meters, or by assigning the Greek prefix centi (meaning hundredth) is one hundredth of a meter. That is a meter is a hundred centimeters, and a kilometer is 1000 thousand meters. The system is useful because not only have we derived the units from physical properties, the different units in the system grow as multiples of ten. So, if we want to work with measuring something small, like an acorn, we describe it with centimeters, and if we work with something large, like the diameter of the earth we use kilometers.

Before the metric system, we measured quantities in terms of feet and pounds, while seconds have always been constant to both the foot-pound system and the metric system. Probably because seconds are derived from the 365 day, twelve month, year and the 24 hour day, which historically developed from wise calendar makers over thousands of years. But the foot makes little sense, only coming perhaps from the hearsay shoe size of a king, and the pound from God knows where. Still further, this system would seem not to be that pragmatic in that there are twelve inches in and foot and three feet in a yard, a far cry from the sensibility of the metric system where all sizes or amounts are a multiple of ten of another.

Europe uses the metric system, like road signs are in kilometers, and shoe sizes in centimeters. We use miles for our road signs and inches for shoe sizes. However, because of the pragmatic nature of the metric system, American and European scientists use the metric system alike. Often, however, because the foot-pound system is still in use in The United States, scientist sometimes find themselves having to go to the effort of converting data they need from feet and pounds to meters and kilograms.

But what if I said not only are both the metric system and the foot-pound system connected to the Universe but to one another in a way that would suggest the evolution of their developments is different than the history we know? That is what I seem to be finding.

The Dance Of The Heavens

It is an extraordinary thing, the dance of the celestial bodies in the heavens that has been observed since ancient times, and found to be describable in terms of the most wondrous aspects of both number and geometry.

Because of the motion of the earth around the sun, the stars will appear in the same position in the sky four minutes earlier each night, as the earth rotates, making them rise and set, so that in the course of the 365 day year they will have made a complete journey around the sky. The Earth goes around the Sun once in this 365 day year, and 365 is close to 360, which are the number of degrees in a circle. We have divided up the circle like this, because 360 is divisible evenly by the degrees in special triangles: The equilateral triangle (60-60-60), the 45-45-90, and the 30-60-90, made by drawing in the altitude of the 60-60-60 triangle.

In turn, the Moon goes around the Earth a little more than twelve times in a year. Twelve is the smallest abundant number. It is evenly divisible by 1, 2, 3, 4, 6, and 12. It nicely orbits the Earth in about 30 days, and we already mentioned 30 divides nicely into 360. So, we have the twelve month year and the 30 day month. We double our smallest abundant number, 12 to get 24, and we divide our day into 24 hours, the day being the time it takes the earth to revolve on its axis.

We then divide each hour into 60 minutes, and each minute into 60 seconds. We already mentioned 60 divides nicely into 360 degrees. We divide each quarter of the moon into 7 days, so that after a month we have close to its approximate orbital period of 30 days.

We have done a good job that began with the ancients, of making the motions of the celestial bodies, which are like a natural clock, to fit them with the key numbers, and geometries. It would seem this is what we have done. It is amazing to me that the earth and moon came about in a way that they move so closely to such a scheme.

Now I find that there is a yin and yang in the Universe of 9/5 and 5/3, respectively from which can be derived what I call the Uranus equation, and, it seems to reconcile the metric system with the foot pound system, and further, though human inventions, show them connected not just to one another, but to the Cosmos.

Do all star systems in the Universe, especially those that host intelligent life, have such an extraordinary structure as ours, perhaps indicating they are in the Universe for a reason, as ours says about us? Manuel claimed he had been all across the Galaxy, and that though different, they did.

The Differential Equation

I have now discovered that Manuel was in England for more than one reason. He was to meet with a Frenchman who was a representative for the Bureau Of Weights And Measures on a mission to change the method of defining that distance called the kilometer.

Apparently it had something to do with his hose collection numbering fifteen hoses and his feeling that God was an idea of which there are two elements. It would seem he carried a great secret that was handed down to him from his ancestors as they passed through the world by foot at times, camel at other times, and caravan at other times completely unknown, starting in India and ending in Spain.

The Differential Equation

I thought it was really time I solve a differential equation. I chose The Uranus Equation.
We now determine given the acceleration in the Uranus equation is 3.3 earth gravities, then what is the initial velocity in the equation, which is the number 3?

$(v) = 3 + 3.3t$ (The Uranus Equation)

$(v) = at = (33/10)t$ where V is velocity, a is acceleration, and t is time.

$3 = (33/10)t$
$(t) = 30/33$

980 cm/s/s = g = the surface gravity of the earth

(980 cm/s/s)(3.3) = 3,234 cm/s/s

(3,234 cm/s/s)(30/33 s) = 2,940 cm/s = v_0 v_0 is the initial velocity

Thus we can write the Uranus equation as:

(v) = 2,940 cm/s + (3,234 cm/s/s)t

This is the differential equation:

(dx) = (2,940 cm/s)dt + (3,234 cm/s/s)t dt

The only thing we lack in solving this is a time for which we can derive a distance. Clearly if the Shaman had some secret, it was in the hose collection he showed me. Of course, then, the time is 15 seconds. The result is

$(x\_0) = (2,940 \text{ cm/s})15 + ((1/2)(3,234) \text{ cm/s/s})15^2 = 44,100 + 363,825 = 407925$ cm

407925 cm/100/1000 = 4.07925 km

Notice the numbers after the decimal are 7925. The seven is the average of the nine and the five, and the nine and five make up the mysterious nine-fifths. The 2 is the number used to average the nine and the five to make the seven. But more incredibly the value is almost perfectly four kilometers. Thus Manuel went to England in my belief not just to spend time with the awesome Sheila Chandra, but to get The Bureau of Weights and Measures to define the length of a kilometer as:

(1/4) (Integral [(2,940 cm/s)dt + (3,234 cm/s/s) t dt] minus 0.07925)

All that divided by 100, then by 1000.

That refers to the integral
100 is 100 cm/m
1000 is 1000 m/km

That integral evaluated from 0 to 15 seconds.

How could he have known so much about the Universe and connected it to the kilometer, when The Metric Committee in modern times defined the kilometer as one ten thousandth of the distance from the North Pole to The Equator?

Maurice Chatelain, who was the director of developing the communications system for the NASA Apollo Missions, may have the answer in his book "Our Cosmic Ancestors". It may be that it was even known longer ago than that day Manuel's ancestors left India. Chatelain writes:

"This is where my interest became really aroused and my work became exciting...All units of measure in the distant past of our civilization had the same basic system in their foundations – all were determined from the exact dimensions of our planet Earth. As incredible as this may sound to the uninitiated, our ancestors derived their feet and inches from the length of degree of latitude or longitude.... How come then that our forefathers back in the Stone Age had values so exact, which we ourselves were only able to obtain after 4 October, 1957, when the Soviet satellite Sputnik started to trace and measure the first orbits around the globe?... We then obtained the exact measurements by observing the irregularities in the orbits of the first artificial satellites in order to calculate the true shape and dimensions or our globe."

Chatelain suggests extraterrestrials, measured the earth by orbiting it in the same way and gave our units of measurement to our past, distant ancestors, 65,000 years ago. The ancestors of Manuel, knew this at least a thousand years ago, and connected it, further, to "The Cosmos". The wind blows in an endless sea of sand. Riding on it is a voice breaking up and disappearing into the air: "La tila, Guadacali".

A Second Look At The Uranus Integral

Uranus Integral From 0 To 15 Seconds

$$\int_0^{15}(2,940cm/s)dt + \int_0^{15}(3,234cm/s/s)tdt = 4.07925km$$

Which can be written in kilometers:

4+(9+5)/2(100) + 9/1000 + 2/(10,000) + 5/100,000

= 4+(0.07)+(0.009)+(0.0002)+(0.00005)

We can also write:

4 km + 70 m + 9 m + 0.2 m + 0.05 m = 4 km + 79.25 m

Or,

4 km + 7000 cm + 900 cm + 20 cm + 5 cm = 4 km + 7,925 cm

Let's convert the part beyond 4 km to feet:

(7,925 cm)(0.0328084 ft/cm) = 260.00657 feet

It is almost exactly 260 feet. The part after the decimal can be arranged as 5,6,7,... It would seem using the Shaman's 15 seconds in The Uranus Integral wants to connect the metric system to feet. It is accurate to about six and a half thousandths of a foot. If we consider the origin of the metric system being in the dimensions of the Earth, and the obscure origins of the foot, this is rather remarkable, especially when one considers the origins of The Uranus Equation.

Now we take the second look:

(260.00657 ft)(12 in/ft) = 3,120.07884 inches

It is almost 3,120 inches! Let's rearrange those numbers. We get:

0, 1, 2, 3,...

Is it that extraterrestrials gave us the inch, the foot, and metric system in ancient times as Maurice Chatellain of NASA suggested in his book, Our Cosmic Ancestors? There has always been a debate between Europe and America as to whether America should adopt the metric system in everyday life. Americans already use it in the sciences. Is there a reason for these different systems of measurement? Were they both meant to be in that they are connected to one another through the Universe and Nature?

The Source Of One Fourth In The Uranus Integral

1 km = 1/(10,000)(1/4)C where C = 2(pi)R_e

C is the circumference of the Earth and R_e is the radius of the Earth.

R_e =6,378 km

We can say:

4 km = (1/10,000)C

Uranus Integral (1/4) (Integral [(2,940 cm/s)dt + (3,234 cm/s/s) t dt] minus 0.07925)

All that divided by 100, then by 1000.

That refers to the integral
100 is 100 cm/m
1000 is 1000 m/km

That integral evaluated from 0 to 15 seconds.

We also call the following The Uranus Integral:

$$\int_0^{15}(2,940cm/s)dt + \int_0^{15}(3,234cm/s/s)tdt = 4.07925km$$

The factor of one fourth enters because the kilometer is defined by the distance from the the pole to the equator, not the circumference of the Earth.

Again, we found 9/5 recurrent in the Universe and Nature. The Gypsy Shaman has indicated that 15 is at the heart of matters. We used 9,5, and 15 to construct the following quadratic:

The area of a rectangle is 15. The length, l, is 9 more than the width multiplied by 5:

$$w^2 + 9w - 3 = 0$$
$$w = 0.32$$

We then made a quadratic in time:

$$15 = \frac{1}{2}gt^2$$
$$t^2 = \frac{30}{g}$$

Because of the result for w, we chose g = 32 feet/sec/sec as opposed to the value in meters g = 9.8 meters/sec/sec

$$t^2 = \frac{30\,ft}{32\,ft/sec/sec}$$
$$t = 0.968\,sec$$

We note that 1-0.968 =0.032

I find that interesting. The Shaman's 15 seconds described meters in the Uranus Integral and the numbers in nine-fifths and their average, not to mention there was the two used to average nine and five, now it is profoundly connected to the 32 of earth surface gravity in feet.

We have another quadratic. The area of ①
a rectangle is 15, after the shaman's
15, the length ℓ is 9 more than the
width multiplied by 5, after our
enigmatic 9/5.

$A = 15 \quad \ell = 5(w+9) = 5w + 45$

$A = \ell w = (5w + 45)w = 5w^2 + 45w$

$15 = 5w^2 + 45w \quad or \quad 5w^2 + 45w - 15 = 0$

~~$\cancel{}$~~ $w^2 + 9w - 3 = 0$

$w^2 + 9w = 3 \qquad w^2 + 9w + \frac{81}{4} = 3 + \frac{81}{4}$

$\left(w + \frac{9}{2}\right)^2 = 3 + \frac{81}{4} = \frac{93}{4}$

$\boxed{\begin{array}{l}(6.32)(46.5) \\ = 14.88 \\ \approx 15 = A\end{array}}$

$w + \frac{9}{2} = \pm\sqrt{\frac{93}{4}} = \pm\frac{\sqrt{93}}{2}$

$w = \pm\frac{\sqrt{93}}{2} - \frac{9}{2} = \pm\frac{\sqrt{93} - 9}{2}$

$\sqrt{93} = 9.64 \qquad w = \frac{9.64 - 9}{2} = \frac{0.64}{2}$

$\boxed{\begin{array}{l}\ell = 5(\sqrt{93} - 9)/2 ~~\cancel{}\\ = (5\sqrt{93} - 45)(2) ~~\cancel{}\\ = \frac{5}{2}\sqrt{93} - \frac{45}{2} ~~\cancel{}\\ \qquad + 45\end{array}}$

$w = -\frac{9.64}{2} - \frac{9}{2} \neq w$

$\frac{0.64}{2} = 0.32 \quad \boxed{\ell = 46.5}$

Pounds And Kilograms

Early on in my work, when I first noticed nine-fifths in the solar system and nature, I was intrigued by the fact that there is nine-fifths of a degree centigrade per every degree of Fahrenheit. I had tried to find the physical connection concerning this, and learned Fahrenheit was constructed so that freezing temperature, or near there, literally the temperature measured by a thermometer in a mixture of salt and ice, perhaps some water, was to be taken as 32 degrees and the upper limit was determined by the human body temperature as measured by a thermometer in an armpit to be a value around 97 or 98. Why these values were chosen I could not find any reason for that was documented. I did post this story to some kind of a discussion group. Later I had posted my discovery of nine-fifths to the Joseph Campbell discussion forum and someone there said nine-fifths was in the equation that converts centigrade to Fahrenheit. I never talked about my earlier work in that area, just said, yes there is nine-fifths degree centigrade per every degree of Fahrenheit. He kept saying you have to add thirty-two. I could not convince him what I was saying was true. Adding the thirty-two is to zero the relationship so one can convert centigrade to Fahrenheit. That is their scales are different: Celsius begins at 0 degrees is freezing and Fahrenheit begins at 32 is freezing temperature. That does not change the fact that there is nine-fifths degree centigrade per every degree of Fahrenheit. He was presenting the formula that converts one temperature to the other.

To the matter at hand: The Gypsy Shaman's 15 seems to reconcile metric distances with feet, both systems use the second to measure time. The Uranus integral is derived from nine-fifths, which we see unifies the metric measurement of temperature with that of the foot-pounds Fahrenheit. That only leaves us needing to find the connection between the metric system's grams and kilograms with the pound, so we have dealt with the matter where mass and weight are concerned. A gram is nicely defined as the amount of water in a cubic centimeter. All I can say, at this point, is that one pound is 453.59 grams, and that makes it exactly 0.59 grams from a whole number. It has something to do with nine-fifths?

Getting A Sense Of The Uranus Equation

Thus we can write the Uranus equation as:

(v) = 2,940 cm/s + (3,234 cm/s/s)t

This is the differential equation:

(dx) = (2,940 cm/s)dt + (3,234 cm/s/s)t dt

Uranus Integral From 0 To 15 Seconds

$$\int_0^{15}(2,940cm/s)dt + \int_0^{15}(3,234cm/s/s)tdt = 4.07925km$$

Mach 1 = 768 mph =1,235 km/hour

That is mach 1 in dry air at 20 degrees C (68 degrees F) at sea level.

If we write, where 1,235 km/hr (mach 1) = 0.343 km/s, then:

34,300 cm/s =2,940 cm/s + (3234 cm/s/s)t

and

t=9.696969697 seconds = 9 23/33 s = 320/33 seconds ~ 9.7 seconds

So, the Uranus equation is a time of 9.7 seconds to reach mach 1. Putting that time in the integral:

(x) = (2,940)(320/33) + 1/2(3234(320/33)^2 = 180,557 cm 1.80557 km ~ 1.8km

Thus in the Uranus Equation we reach mach one in about 9.7 seconds after traveling a distance of about 1.8 kilometers. Let's convert that to miles:

(1.8km)(0.621371mi/km)=1.118478 miles

Once again we see nine-fifths unifying the foot-pound system to the metric system in that 1.8 is nine-fifths and a mile is about 1.8 kilometers. That is if we think of the mile as part of the foot-pound system.

The Uranus Integral Is Intriguing

The Uranus Integral reaches mach one in 1.8 kilometers, which is the amount of kilometers in a mile and is the 9/5 that occurs in Nature and the Universe, not to mention that it unifies pi and golden ratio and pi and Euler's number e. We have already shown how the Uranus Integral describes, kilometers, nine-fifths and feet and inches, when evaluated at the Gypsy Shaman's Mysterious 15 in the form of 15 seconds. Fifteen is the degrees the Earth rotates through in an hour. We have shown how nine-fifths unifies grams and pounds and centigrade and fahrenheit.

Thus, not only does the Uranus Integral describe kilometers, it describes miles. It is one compact statement that embodies everything. Again we see the Shaman's 15 unifying two systems of measurement in that there are fifteen degrees in one hour of right ascension.

More 9/5 In Our Units Of Measurement

It would seem the temperature scale is connected to 9/5 in more than the ways I wrote about in Files U And E to add it too the long list of nine-fifths in our units of measurement that I have already found and its connection to pi and the golden ratio and pi and euler's number and key occurrences in nature. Room temperature is 68 to 77 according to wikipedia. The average (68 + 77) = 72.5. The seven and two add up to the nine in nine-fifths and the five after the decimal is the five in nine-fifths.

Ian Beardsley
September 13, 2013

A New System Of Measurement Seems To Suggest Itself

Concerning my work "Files U And E" and "Notes 01 For Files U And E" we should be able to create a system of units for mass, length and time that connects not just the metric system with the foot pound system, but that connects with Nature and the Universe from flower petal arrangements to the physical characteristics of the Earth, planets, and stars. That is what I am finding. I have even suggested that the way we constructed our measuring systems came about from extraterrestrial influence as it seems to be too dynamic to have evolved through random forces of history. I even seem to have found a message from extraterrestrials embedded in our physics indicating that this is what extraterrestrials are hinting we should do. I have even found that that message seems to come from the same region of space as SETI's Wow! Signal.

Ian Beardsley
September 14, 2013

When we say nine-fifths is in the molar mass of gold to the molar mass of silver and in the solar radius to the earth-moon separation, we might want to point out that it is appropriate because the sun is gold in color and the moon is silver in color.

When I say that Maurice Chatelain may have the answer when he says extraterrestrials may have gave us our units of measurement in ancient times, I mean mostly what I said at the beginning of my work Files U And E, that extraterrestrials left their thumbprint in our physics. What that means is that they have been among us on scientific committees and in other places, mostly in the past five hundred years, influencing the outcome of human history, i.e. guiding us towards the definitions of the units of measurements we have today. The definitions just have too much relation to one another and nature in ways that were not intended or even known about. That is what I am finding.

In Files U And E by Ian Beardsley, we opened with a story about a Gypsy Shaman, Manuel, which lead to the realization that not only was nine-fifths at the crux of the Universe or Nature, but mathematical constants such as pi, the golden ratio, and Euler's number, e. Further we found this nine-fifths was not just in our systems of units of measurement, connecting not just one to the other, but also both to the Universe. In so far as the values we assigned to our units of measurement should be a random collection of numbers, as their history seems to be one of random evolution, and they are not, we suggested extraterrestrials left their thumbprint in our physics, that is had a hand in its development. We also found what seems to be a message from extraterrestrials embedded in our physics and we were able to calculate the direction in the sky that is its origin. It incredibly turned out to be in the same region of the sky as the origin of the Wow! Signal, which was the one signal SETI (The Search For Extraterrestrial Intelligence) ever received that had all the characteristics they were looking for in a message from extraterrestrials.

After discovering the properties of 9/5 in The Universe And Nature, and noting that it was in the five-fold symmetry (regular pentagon) that is typical of life, we called it the yin of the Universe and derived the yang of the Universe from the six-fold symmetry we find in the physical Universe (regular hexagon). It was five-thirds. Thus we created a taoistic Universe (The Gypsy Shaman said "God is an idea, there are two elements").

This in turn lead to our Neptune and Uranus Equations, which we guess suggests some connection of the planets Neptune and Uranus to human destiny.

My Cosmic Archaeology is based on the idea that 9/5 is the Yin of the Universe, 5/3 is the Yang of the Universe, and 15 (Manuel's number, M) is the magic number. With these three numbers we construct a Universe that seems strikingly more than coincidental. Also of interest to this story is The AE-35 Antenna. I wrote:

AE-35

I wrote a short story last night, called Gypsy Shamanism and the Universe about the AE-35 unit, which is the unit in the movie and book 2001: A Space Odyssey that HAL reports will fail and discontinue communication to Earth. I decided to read the passage dealing with the event in 2001 and HAL, the ship computer, reports it will fail in within 72 hours. Strange, because Venus is the source of 7.2 in my Neptune equation and represents failure, where Mars represents success.

Did extraterrestrials really anticipate, when they sent us in 1977 a transmission from the constellation Sagittarius, that lasted 72 seconds (SETI's Wow Signal) that, a Gypsy Shaman, around 1990, would make me into The AE-35 Antenna of 2001: A Space Odyssey that the ship computer, HAL, would say would fail within 72 hours, the same 72 of the Wow Signal? Did they know the transmissions I would receive from the Gypsy Shaman, Manuel, would lead not just to the discovery of a message embedded in our physics, but a system of three equations that would point to the same region of space as the Wow Signal, the constellation Sagittarius, for the origin of my message?

This is the same 72 that is in the 0.72 astronomical units of Venus from the Sun. It is the same 72 that is the number of years for one degree of precession of the Earth's Equinoxes.

Putting myself to thinking about the meaning of the AE-35 antenna in 2001: A Space Odyssey, I would say it is, in the film, the turn around point for humankind between being a being that depends on technology and a being who is liberated from technology; between being still an ape, although one that is spacefaring, and the starchild. How is this pertinent to my connection between an extraterrestrial message, the Wow! Signal (duration 72 seconds) and the time estimate by HAL of at least 72 hours before failure of the AE-35 unit? My 7.2 in my Neptune Equation and my discovery of what seems to be a message embedded in our physics from extraterrestrials from the same region of space as the Wow! Signal (Sagittarius)? It would seem to indicate, if my interpretations of the math are correct, that extraterrestrials are telling us we are at a turn around point ourselves, now, or that they are going to invite us into stellar society soon.

The SETI extraterrestrial signal called the Wow! Signal was detected on August 15, 1977. Not only was it on the 15th day of August but in 1977. The 7 and the 7 add up to fourteen, add the one and you have 15. Nine is the Nine in the nine-fifths that is at the basis of the strange enigma my work is about and the fifteens are symbolic of the Gypsy Shaman's hose collection of 15 hoses that started the project. My guess is this all has everything to do with Manuel and his ideas about the AE-35 antenna. To know what I am talking about, read my work The Uranus Enigma by opening the document below. Also, The one and the 9 in 1977 add up to 10. That is a one and a zero. Add the one to the two sevens and you have 15 again. It would seem the extraterrestrials covered all angles.

A New System Of Measurement Seems To Suggest Itself

Concerning my work "Files U And E" and "Notes 01 For Files U And E" we should be able to create a system of units for mass, length and time that connects not just the metric system with the foot pound system, but that connects with Nature and the Universe from flower petal arrangements to the physical characteristics of the Earth, planets, and stars. That is what I am finding. I have even suggested that the way we constructed our measuring systems came about from extraterrestrial influence as it seems to be too dynamic to have evolved through random forces of history. I even seem to have found a message from extraterrestrials embedded in our physics indicating that this is what extraterrestrials are hinting we should do. I have even found that that message seems to come from the same region of space as SETI's Wow! Signal.

Ian Beardsley
September 14, 2013

We have taken Manuel's Integral and used the new suggested length of a meter, and found that it makes Earth Gravity round closely to an even 10 meters per second squared. This is important because not only do we have a base ten number system – probably because we have ten fingers to count on – but the metric system is based on multiples of ten. The other interesting aspect is that this new definition for the meter (m) that we will call new meters (nm), gives a period of two seconds for a pendulum of string length one new meter and therefore a swing from left to right or right to left of one second. We have further connected new meters to the ratio of gold to silver in molar masses, the ratio of the solar radius to the orbital distance of the moon, and much more in nature if you read "Files U and E" by myself. (Not to mention the five-fold symmetry typical to life, like a five-petal flower).

We begin with Manuel's Integral:

$$\int_0^{15}(2{,}940\,cm/s)dt + \int_0^{15}(3{,}234\,cm/s/s)t\,dt = 4.07925\,km$$

We make kilometers smaller as suggested by the integral:

4km/4.07925 = 0.980572409

And we say (rounding it to three places after the decimal),

0.981 cm/1.000 ncm

Where ncm means new centimeters.

We then notice the new centimeter is Earth Gravity (981 cm/s/s) reduced exactly by a factor of 1000 giving earth gravity in new centimeters a value of exactly 1000 new centimeters per second squared:

(981 cm/s/s)(ncm)/0.981 cm)=1000 ncm/s/s

Converting this to new meters (nm) we have for earth gravity (g):

(g) = 10 nm/s/s

That is with new meters, earth gravity is exactly ten new meters per second squared. This was derived from Manuel's Integral, which we have shown in my work Files U And E to have a profound connection to nature. It reaches mach 1 at sea level and 68 degrees Fahrenheit in 1.8 kilometers, which is 9/5 that occurs in nature in gold, silver, sun, moon, Earth, Jupiter Saturn, and that unifies pi with the golden ratio and pi with euler's number, e. Now it seems to yield a new meter for which earth gravity is ten, and base ten is our number system, and that on which the metric system is based. Now let us look at the period of pendulum with a string length of one new meter:

$$T = 2\pi\sqrt{\frac{nm}{10\,nm/s/s}} = 1.986917653 \approx 2\sec onds$$

A string of new meter length rounds to a period of two seconds, which means it has a swing from left to right or from right to left of one second.

Thus we define new meters as 0.081 meters and seconds stay the same.

Ian Beardsley
September 16, 2013
Creating A System Of Units

The whole idea is to redefine the units of measurement that are at the foundation of physics – mass, length, and time – such that they have a profound connection with nature. It has been suggested in my work, Files U And E and their notes, that such an endeavor should be undertaken. We have already dealt with length in our definition of new meters (nm) and we have suggested that the units for time should stay the same (seconds) because they are already profoundly connected to nature in their connection to the rotational period of the earth and its orbital period, if not to other things. We are then left with dealing with mass, (the pound and kilogram, grams,... etc.)

We consider then, the time it takes to reach mach one at the standard conditions (68 degrees Fahrenheit and the pressure at sea level) which is 320/33 seconds to use a close fractional approximation, This gives:

(v) = (2,940 cm/s) + (3,234 cm/s/s)(320/33 s) = 29.40 m/s + 313.6 m/s = 343 m/s

We see this approach not only gives a whole number velocity, but the digits in the value add up to the ten of the base ten counting system. This would indicate that it is a move in the right direction, if not an example of what we should not do. We find the momentum of such a velocity for one kilogram and it is the number itself:

(1 kg)(343 m/s) = 343 Newton seconds

We seek a velocity in nature so we can find a new value for kilograms we can call new kilograms (nkg).

That is,

(mass)(velocity) = 343 Newton seconds

Then we look at the functionality of such a value for mass, in nature.

Ian Beardsley
November 7, 2013

Is The Extraterrestrial Message Real?

I was lead in my research to think there was good reason we should try to create a system of units for measurement of physical reality that not only related the foot-pound system to the metric system but that connected to Nature in a profound way, or, if not, use the connection between the two systems to aid us in finding one that is more profoundly connected to nature than anything we have so far.

Interestingly, I had suggested we should not change the unit of time, one second, because I believed it was already connected to nature in a profound way, as I outlined in my book SETI: Another Signal In Sagittarius in the chapter at the beginning titled The Dance Of The Heavens.

I find that interesting because that profound connection lay in the connection of the three special triangles to the 365 day year in that 365 is close to the 360 degrees in a circle, which means each day of the year represent about one degree of motion of the Earth around the Sun.

I now realize that as well, the importance of my hypothetical message from extraterrestrials is in the same idea: That, it connects the Earth year to the three special triangles. Does this support the idea that the hypothetical message from extraterrestrials is real and that we should really look for a system of measurement that is more profoundly connected with nature?

Ian Beardsley
December 3, 2013

New Seconds

Let's try to make a new unit for the duration of a second, and call it new seconds (ns). We derive it such that the earth moves through one degree in a day.

(31,557,600 seconds/year)/(360 days) = 87,660 new seconds/day

(24 hours)(60 min)(60 seconds) = 86,400 second/ day

87,660 ns/86,400 sec = 1.014583333 ns/sec = 487/480 ns/sec

Or,

0.985626283 sec/ns = 480/487 sec/ns

We found the period of a pendulum in new meters (nm) was

T = 1.986917653 seconds

Convert that to new seconds (ns) and we get:

(1.986917653 seconds)(1.0145883333 ns/sec) = 2.015893529 new seconds

That means if we use new meters and new seconds the period of a pendulum with length one new meter has a period of almost exactly two new seconds (to a hundredth of a second) which means one swing from left to right or from right to left has a period almost exactly one new second.

Randomly Evolved Units Not So Random In Nature

We have found 9/5 in both the metric system and foot-pound system, temperature in centigrade and fahrenheit and to be connected to nature and to unify pi and the golden ratio and pi and euler's number. Earth gravity (g) in the metric system is one more to add to the list:

981 cm/s^2

It has the nine of nine-fifths, the one and eight that make up the 1.8 that is 9/5, and the eight minus the one (7) is the average of nine-fifths.

We have suggested that extraterrestrials influenced the events that lead to our defining of units. Read "Files U And E" by Ian Beardsley.

Ian Beardsley
September 12, 2013

Why try to find another system of units that is more profoundly connected to Nature and The Universe?

Because then using it in the course of scientific research, a grand design for Nature and the Universe could plop out on our laps as a side effect. I have already found that our current systems of units, both metric and foot-pound, are profoundly connected to not just the Universe and Nature, but to one another.

These systems have a history of random evolution from defining, for example, weight by the weight of a coin and particular king minted, to committees debating and voting on how to create the metric system. I have therefore suggested that extraterrestrials have somehow influenced the outcome of their complicated evolutions, perhaps to use us as an experiment for their own ends, or to help us.

However, I have suggested one other mechanism for the dynamic structures they have, and that is that the way we define units through a random evolution could have a tendency towards the numbers I am finding which are built around 9, 5, and 15. It has already been discovered in a mathematical theory, which has been verified, the name eludes me, that numbers obtained in taking data with respect to Nature, will typically have leading digits more recurrent than others. I believe the number 1 is mostly likely to occur, and I remember that 9 is the least likely. Funny because 9 is precisely the number I find most in the definitions of our units, like 9/5 degrees centigrade per unit Fahrenheight and 981 cm per sec squared for earth gravity. But then for every Camus, there is a Sartre, and I might just be seeing the other side of the coin.

I have developed a Gypsy Cosmology based around the three numbers 9/5, 5/3. and 15. The number 15 came to me by way of a Gypsy Shaman called Manuel. Now I have found a connection to the earth climate. Let me first briefly discuss some of the findings, then I will present the connection to climate, then I will present one page that goes into some of the deeper elements concerning the Gypsy Cosmology.

When we say nine-fifths is in the molar mass of gold to the molar mass of silver and in the solar radius to the earth-moon separation, we might want to point out that it is appropriate because the sun is gold in color and the moon is silver in color.

I now further go on to say that this nine-fifths unifies the two most important ratios in mathematics pi and the golden ratio (phi), in that

pi + phi = 3.141 + 1.618 = 4.759

Because the numbers after the decimal in the sum (the important part) are 5 and 9 and 7, the average of 5 and nine.
It would seem 9/5 unifies euler's number, e, and pi, as well:

(pi) + (e) = 3.141 + 2.718 = 5.859

The second number after the decimal is the 5 in nine-fifths, and the third number after the decimal is the 9 in nine-fifths. The first number after the decimal is eight. This is significant because the 8 is the 8 in 1.8, which is 9/5 divided out. The one will take you all the way around a circle, what is left is 0.8.

To read the full thesis, which proposes extraterrestrials left their thumbprint in out physics, even a message, and for which I calculate its origin to be in the same place, Sagittarius, as the SETI Wow! Signal, see my paper at:

http://issuu.com/eanbardsley/docs/more_online

Called "More That Can Be Said".

Ian Beardsley
February 27, 2014

Manuel's Number And Nine-fifths Connected to Climate

The solar luminosity is:

$$L_0 = 3.9 \times 10^{26} J/s$$

The average distance of the Earth from the Sun is:

$$1.5 \times 10^{11} m$$

Therefore the solar constant is:

$$S_0 = \frac{3.9 \times 10^{26}}{4\pi(1.5 \times 10^{11})^2} = 1{,}370 watts/meter^2$$

That is the amount of energy per second per square meter hitting the Earth.

The radiation, F, is proportional to the temperature, T to the fourth power, and equal by the Stefan-Boltzman constant, sigma:

$$F = \sigma T^4$$
$$\sigma = 5.67 \times 10^{-8} Wm^{-2}K^{-4}$$

This gives the temperature, T, at the top of the Sun's photosphere is:

T=6,000 degrees Kelvin

The planetary albedo, a, is the amount of radiation from the Sun that the Earth reflects back into space which is 30%. Therefore a=0.3 is the planetary albedo. Therefore the Earth receives 70% of the Sun's light, or, in other words:

$$\frac{S_0(1-a)\pi r^2}{4\pi r^2} = \frac{S_0}{4}(1-a) = \sigma T^4$$
$$T = 255K = -18°C$$

That is the temperature the Earth would be if it had no atmosphere, minus eighteen degrees centigrade. The observed average temperature is:

$$T = 15°C$$

Fifteen degrees centigrade. The without an atmosphere temperature, the minus 18 degrees centigrade, is the one and eight of 1.8 that is the Nine-Fifths upon which we have built our Gypsy Cosmology. The fifteen Degrees is the Gypsy's 15.

The Genesis Hypothesis

It is my guess that The Genesis Project – making a life sustaining planet – would be attainable by using a planet in orbit around a star of a luminosity that is absolute magnitude zero at a distance from it that is the same as the distance between the planet Saturn and the Sun. I call this "The Genesis Hypothesis".

The hints hidden in Nature that leads me to this guess are:

1. The energy required to move the earth from the sun to its orbit of one astronomical unit is about 94.2 solar luminosities radiated over a year, a number close to the number of naturally occurring elements, which are 92. 92 solar luminosities are close to the luminosity of an absolute magnitude zero star. There are precisely 85.525. If we average this with 100 solar luminosities we have 92.7625 solar luminosities, very close to the number of naturally occurring elements.
2. Saturn has the most prominent ring system in the solar system. It is about 10 times further from the sun than the Earth is from the sun, and if an absolute magnitude zero star is about 100 solar luminosities, by the inverse square law the habitable zone of an absolute magnitude zero star is the same distance from that star as Saturn is from the Sun.
3. I found absolute magnitude zero stars have a connection with an enigmatic occurrence of 3.7 in the Universe.

Concerting number 3, the latter point, I had written about through my character Leonard in The Secret Of Neptune:

Leonard referred to his periodic table of the elements and noticed that the ratio of Gold (chemical symbol Au) to Iron (chemical symbol Fe) was 3.5. Why was this important? He figured it was important because the luminosity of a star increases exponentially as the mass of a star, and that exponent was 3.5.

Gold has been considered the most precious metal since before the Ancient Egyptians Buried their kings in tombs with treasures made of the stuff, to today as the most precious of ceremonial jewelries, and the use of Iron marked the beginning of an age for humans that continues to this day. He noticed if he let an absolute magnitude zero star, which he had shown was important, be rounded to 100 solar luminosities, which put its habitable zone at Saturn orbit around the sun, and used that number in the equation, it would have a mass of 3.7 suns. He found 3.7 was important in his following discovery:

[(volume of saturn)/(volume of Jupiter)](volume of mars)

=(mercury radius)(earth radius)^2

=[(venus orbit/earth orbit)(earth radius)]^3

=[(mercury orbit)/(earth orbit)](earth radius)^3

=0.37 cubic earth radii

0.37 can be converted to 3.7 by multiplying it by the ratio of earth mass to mars mass because it is close to ten.

(earth radius)/(moon radius)=

4(degrees in a circle)(moon distance)/(sun distance)

= 3.7

There are about as many days in a year as degrees in a circle.

… And therefore absolute magnitude zero stars (blue stars) were once again relevant to the human story.

I had further written a treatise on Blue Stars, as follows"

In order to develop a comprehensive exopolitics (politics between the spacefaring worlds) one has to categorize the different kinds of stars and explain their relevance since they are those things, which warm the worlds. I originally thought blue stars were of prime importance, but it may be all stars (which we will divide into four categories: red, yellow, blue, white) are important in different ways. I here begin the process, starting with blue stars.

My calculations concerning absolute magnitude zero stars (Blue Stars) are as follows:

Chapter 8: Leonard Calculates A Functional Star System

Entry: Topic: Blue Stars

We will find the energy required to move the earth of mass m, from the surface of the sun under its gravity, to its current orbit of one astronomical unit, the sun of mass M.

The energy given to move something against gravity is given by the Work, W:

Work = W = F ds

F is Force. In the case of the sun and earth, we use Newton's Universal Law of Gravity:

$$F = \frac{GMm}{r^2}$$

We integrate from R to r (from the radius of the sun to the orbit of earth):

$$W = \int F \cdot dr = -\frac{GMm}{r}$$

G is the universal constant of gravity.

G = 6.672E-08 in dyn-centimeter squared-per second squared
And, M = 1.989E33 grams
And m = 5.976E27 grams
And R =6.9599E10 centimeters
And r = 1.495979E13 centimeters

Evaluating the integral for work, W:

$GMm = (6.672E-08)(1.989E33)(5.976E27)$

Which is equal to 7.93E53

(7.93E53)/(6.9599E10) = 1.14E43 ergs

(7.93E53)/(1.495979E13) = 5.30E40 ergs

1.14E43 – 5.30E40 = 1.13E43 ergs

1 erg = 1E-07 joules

The final answer is 1.13E36 Joules of energy to move earth from the sun to its orbit.

We now calculate how many times brighter the sun has to be to equal the annual output of energy (luminosity) of what we just calculated, the energy to move the earth from the surface of the sun to its current orbit.

Seconds in a year:

(365 days)(24 hours)(60 minutes)(60 seconds) = 3.15E7

Solar luminosity is: 3.826E26 J/s

(3.826E26 J/s)(3.15E7 s) = 1.21E34 J/yr

(1.2E34)x = 1.13E36 J

And, x = 94.2

The sun must be 94.2 times as luminous for its annual output in energy to equal the energy required to move the earth from its surface to its current orbit.

We calculated that an object has to be about 94 times as luminous as the sun to have an annual output in energy over a year, that equals the energy required to move the earth from the surface of the sun to its orbit, and there are 92 naturally occurring elements as of yet, a number close to the factor of 94.

Let us round that luminosity to 94 solar luminosities, and calculate the absolute magnitude of such an object.

$2.512^x = 94$

x log 2.512 = log 94

0.4x = 1.973

x = 4.93

The sun has an absolute magnitude of 4.83 which is about five. The absolute magnitude of our object is 4.93 steps in magnitude brighter than the sun, which is about five magnitudes brighter than the sun. Five steps brighter is five minus five, which gives our object an absolute magnitude of zero. Thus the system of the ancients has a zero magnitude related so elegantly to the number of the naturally occurring elements and the energy required to move the earth from the sun to earth orbit.

A zero absolute magnitude object is a star of spectral class B on the main sequence. That is its color is blue.

The habitable zone of a planet is the orbital distance from a star that allows water to exist as a liquid. Since our zero magnitude star is about 94 times more luminous than the sun, as I have calculated it, the habitable zone for a planet around this zero magnitude star is further from it than the earth is from the sun. Luminosity, or the amount of light given off by a star decreases as the square of the distance from the star (see the inverse square law). Since this magnitude zero star is about 94 times brighter than the sun, and there is some play in the habitable zone, we can, for ease of calculation, and elegance, say our star is 100 times more luminous than the sun. Since ten squared is 100 the habitable zone around our blue, spectral class B, zero magnitude star is 10 astronomical units from it. This puts the orbit of a planet around the star at the same distance Saturn is from the sun.

The stars on the main sequence form an S shaped curve where luminosity increases with mass. The sun is exactly in the middle of this curve, an average, yellow star of spectral class G. The relationship from a sample of many stars shows that mass, M is related to luminosity, L as follows:

$L = M^{3.5}$

We can find the mass of our zero magnitude star with this:
$100 = M^{3.5}$

log 100 = 3.5 log M

(2)/(3.5) = log M

0.57 = log M

$M = 10^{0.57} = 3.7$

Our zero magnitude, blue, spectral class B star is 3.7 solar masses.

The mass of the sun in grams is:

1.989E33 g

This means our zero magnitude blue, spectral class B star has a mass of:

(1.989E33 g)(3.7 solar masses) = 7.36E33 grams

We can use Kepler's third law of planetary motion for circular orbits to calculate the year of a planet in circular orbit around our zero magnitude, spectral class B, blue star. It is:

$$T = 2\pi\sqrt{\frac{r^3}{GM}}$$

Where T is the orbital period, or year of the planet in question, and r is its orbital distance from the star, G is the universal constant of gravitation, and M is the mass of the star the body is orbiting.

$2\pi = 2(3.141) = 6.282$

r = 10 astronomical units

1 astronomical unit is 1.5E13 cm

(1.5E13 cm)(10 astronomical units) = (1.5E14 cm)

$r^3 = 3.375E42$

G=6.672E-8 (dyn cm squared)/(g squared)

$$\frac{r^3}{GM} = \frac{3.375E42}{(6.672E-8)(7.36E33)} = 6.87E15$$

$\overline{)6.87E15} = 8.3E7$ seconds

(8.3E7 seconds)(6.282) = 5.2E8 seconds

(5.2E8 seconds)/(60 seconds) = 8.67E6 minutes

(8.67E6 minutes)/(60 minutes) = 1.445E5 hours

(1.445E5 hours)/(24 hours) = 6.025E3 days

(6.02E3E3 days)/(365.25 days) = 16.48 earth years

The orbit of the planet around our zero magnitude, spectral class B, blue star is 16.48 earth years. The closest orbital period to this in our solar system is that of Jupiter at 11.86 years. That of Saturn is 29.5 earth years.

Stars are approximate blackbody radiators, where a blackbody is that which absorbs all incoming radiation, and emits the maximum amount of radiation for its temperature. We ask, according the laws of blackbodies, given the luminosity of an absolute magnitude zero star, and its surface temperature, what is its radius.

The color index of a star (B-V) 100 times more luminous than the sun is on an H-R diagram, -0.12, which corresponds to many absolute zero magnitude zero stars, and surface temperature of 13,000 degrees Kelvin. Temperature of a star is related to its power radiated per unit surface by the Stefan-Boltzmann law:

$$\frac{R}{R_s} = \left(\frac{T_s}{T}\right)^2 \left(\frac{L}{L_s}\right)^{\frac{1}{2}}$$

Where R is the radius of the star, R subscript S is the radius of the sun, T subscript S is the temperature of the sun in degrees Kelvin, T is the temperature of star in degrees Kelvin, L is the luminosity of the star, and L subscript S is the luminosity of the sun.

We have:

$(5800/13,000)^2(2.512^{4.83})^{\frac{1}{2}} =$

$(0.199)(9.2) = 1.8$ solar radii

We have used the temperature of the sun 5800 degrees Kelvin and $4.83 - 0 = 4.83$ where 4.83 is the absolute magnitude of the sun.

Our absolute magnitude zero main sequence star is 1.8 times larger than the sun.

We have said that a star on the main sequence of 100 solar luminosities is close to an absolute magnitude zero star. Just how close is it? We use:

E1/E2 = 2.512^-(0-4.83) = 2.512^4.83

Where E1 is the luminosity of the star, and E2 is the luminosity of the sun, 4.83 is the absolute magnitude of the sun. This gives:

E1 = 85.525E2

Or in other words, an absolute magnitude zero star is 85.525 solar luminosities. This is very close to our 100 solar luminosities considering a star can be more than a million solar luminosities

I believe zero magnitude stars on the main sequence are important, if not off the main sequence, in that they are connected to our star, the sun, through the number of naturally occurring elements, the earth orbital period, and the amount of energy it would take to move the earth from the sun to its orbit. Such stars, or plus or minus one magnitude of brightness, are of spectral class A to spectral class B. They are important because the sun is important, as it is a star in which a life bearing planet, the earth, has formed.

I believe Saturn is important, because it is at an orbital distance from the sun that is in my projected habitable zone for a zero magnitude star, on the main sequence. Because of this, I feel it is no coincidence that Saturn has rings. What we have here, in this work, is a clue to the secret of origins, in particular to the structure that allows for life.

The absolute magnitude of a star is its luminosity at a set distance from the observer, which has been set at 10 parsecs. One parsec is 3.26 light years, where a light year is the distance light travels in one year, or in the time it takes the earth to make one revolution around the sun, in other words. A parsec is also the parallax of a star measured by displacing oneself by an astronomical unit, where an astronomical unit is the average distance of the earth from the sun, in its nearly circular orbit. This corresponds to a parallax angle of one arc second.

My belief is if you want to get at what nature is, don't worry about being right on in many aspects of work, because in art a circle unclosed is a circle closed by the concept in art they call closure, a nature of the way the human eye works with the mind. Is what we are seeing here is the rough sketch of nature, the idea.

We have said an absolute magnitude zero star has 85.525 solar luminosities and that it is convenient to call it 100 solar luminosities at times. We average the two values:

$(85.525+100)/2 = 92.7625$

There are exactly 92 naturally occurring elements. This average is extremely close to that. We use this fact in our genesis project.

Chapter 23: Nature Speaks Of Human Destiny

Indeed our destiny seems to be outlined by the structure of nature, and I wouldn't be surprised it has extraterrestrial ties. I wrote:

The Nine-Fifths Mystery Begins to Expand: Alloys

This is really exciting, and it shows my reasoning may be correct, to think that human activities have a divine connection to the cosmos. I had found that Saturn (second largest planet) orbit to Jupiter (first largest planet) orbit was 9 to 5 putting the Earth at one unit from the Sun. Amazingly I found the solar radius to the lunar orbit was the same thing. What is more the most precious of the metals in molar mass, gold to silver was 9 to 5 as well.

It would seem as early as 6000 BC the Sumerians were pounding gold and silver out to sculpt things until they ran out. They were doing this with copper as well, another soft, malleable metal, until they ran out. After running out they had to learn to separate the ores of these metals from other substances. Eventually they invented kilns that allowed them to melt the stuff and mix it with tin, to create the first alloy, which we call bronze. This began an age called The Bronze Age, as early as 3500 BC. It was the beginning of metallurgy.

The Sumerians invented the kilns in the first place to fire clay. Today, making Commercial Bronze uses zinc as the alloying metal. So I guessed, just as Gold to Silver is 9 to 5, so should tin to zinc be the same. This is exactly what happened:

$Au/Ag = 196.97/107.87 = 1.82599 \sim 9/5 = 1.8$ (gold to silver)

$Sn/Zn = 118.71/65.39 = 1.81541 \sim 9/5 = 1.8$ (tin to zinc)

I had also found that nine-fifths unifies pi and the golden ratio, and that it is in the rotation of petals around a five petal flower, a most popular arrangement (five fold symmetry is very typical to life). See "Astronomia Flamenca".

For well over a hundred thousand years, humans wander and gather, then all of the sudden one tribe, the Sumerians settle down along the Tigris and Euphrates in Mesopotamia (Iraq) and build brick houses, invent agriculture, invent writing, and metallurgy, medicine, coins, accounting, mathematics and civilization in general. It has been a considered a great mystery why, and even how, they did this, almost overnight on an archaeological time scale. That they would do it seems to be written in the stars, and I wouldn't be the first to suggest it was extraterrestrials of intelligence far beyond that of the human (homo sapien) that had a hand in all of this.

My discovery of nine-fifths in nature lead to my neptune equation and to my uranus equation. Nine-fifths I found was the yin for which yang was five-thirds.

What does this mean? Well if Gold to Silver is 1.8, and tin to zinc is 1.8 (Bronze Age), then should not the next number in the nine-fifths sequence, which is 3.6, represent The Iron Age. Well it does. As it turns out, gold (Au) to Iron Fe is 3.5, dividing out molar masses, as we stated in The Genesis Hypothesis.

After the Iron Age, one might consider it is now, and is the Silicon Age, because silicon allows for integrated circuitry. We could call it the age of electronics, perhaps better the age of digital electronics. Silicon should then be the third term in the nine-fifths sequence, which is 5.4. But we would have to divide it into something. One would guess it would be Europium, because it has no biological function, just a use for electronics in that it is phosphorescent (glows). Well just as I guessed, Europium (Eu) divided by Silicon (Si) is 5.4:

$$Eu/Si = 151.97/28.09 = 5.41011036 \sim 5.4$$

It is almost exact!

The Age of Interstellar Travel

Tin to zinc was the 1.8 of The Bronze Age. Gold to Iron was the 3.6 of The Iron Age. Europium to Silicon was the 5.4 of the Digital Electronics Age. The next number in the nine-fifths sequence is 7.2. We assume that is The Age of Hyperspatial Travel. In other words the age of the development of a starship. There has been some indication this would be powered by the recent addition to The Periodic Table Of The Elements, Element 115 (atomic mass 288). The element was recently synthesized in 2003. To determine the element that divides into it to make 7.2, we write:

$$288/x = 7.2$$

$$7.2x = 288$$

$$288/7.2 = 40 = Calcium = 40.08$$

The element Calcium. I hoped to find something that would be a clue to making hyperdrive (a star engine). But Calcium is the most abundant metal in many animals, mineralizing to make bone, teeth, and shells. It is also the fifth most abundant element in the Earth's crust by mass.

The Coupling Of Mars And Earth: Manuel Speaks To Me By Telephone From Spain To California

Manuel: "Ian, I can't get over what you said earlier that two of the Pythagorean Solids have volumes, one of nine tetrahedra and the other of five tetrahedra. There are only five Pythagorean solids and two of them have volumes the same numbers as the distance to Jupiter and Saturn in their closest approaches to the sun if the earth is at one unit from the sun! This means the tetrahedron represents the earth because it has a volume of one. This is interesting because the solar radius to the earth-moon distance is 9 to 5, and 9 to 5 is the molar mass of gold to the molar mass of silver."

"Why is this so incredible? It is incredible because the tetrahedron also has four faces and Mars is the fourth planet. It is appropriate that the tetrahedron couples earth and mars, further, in that it is four faced (mars is the fourth planet from the sun), and each face has three vertices (earth is the third planet from the sun). You may not think that is incredible but consider categorizing the planets with the smallest abundant number, twelve. Smallest abundant number means it is the most divisible for its size. Twelve is divisible by 1, 2, 3, 4, and 6 a whole number of times. The sum of these numbers is 16, which is greater than 12. Mars is the fourth planet out from the sun. Twelve divided by four is three, and the earth is the third planet out from the sun. Thus once again the earth and Mars are coupled."

"I can couple them again. Elements are atoms, and the numbers of particles that constitute them determine the element, and its relative mass, or "molar mass" as it is called. The atmosphere of Mars is mostly CO_2, about 95 percent. The earth atmosphere is more than 90 percent oxygen and nitrogen. I calculate CO_2 to have a molar mass of 44.01 and earth air to have a molar mass of 28.5756. The Mars orbital distance compared to the earth orbital distance is about 1.5. The molar mass of CO_2 compared to the molar mass of air is about 1.5. That is the mars-sun distance is to the earth-sun distance as Martian air is to earth air."

"It goes further than that. If the earth surface gravity is 1, that of Mars is 0.380. That is a ratio of 2.63. The molar mass of oxygen gas (O_2) what we breath, is 32.00. The molar mass of carbon, the basis of life as we know it, is 12.01. That is a ratio of 2.66. That says it takes about the same energy to lift a mole of carbon on earth, as it does to lift a mole of oxygen (O_2) on mars the same distance."
"Earth and Mars are yet further coupled:"

"The earth atmosphere was once mostly carbon dioxide (CO_2) like the mars atmosphere is today, until plant life came along and started converting the carbon dioxide into oxygen, using light from the sun and in the process making the most fundamental sugar, glucose (which is at the bottom of the food chain). A process called photosynthesis."

"Glucose is C6H12O6. I calculate glucose has a molar mass of 180.16 and water, which covers most of the surface of the earth, chemical formula H2O, I calculate has a molar mass of 18.016."

"180.16 divided by 18.016 is about 10 and the earth is about 10 times as massive as mars. That is, glucose is to water as the earth mass is to the mass of mars."

"You are probably wondering why I am so amazed that Mars and earth are so coupled empirically. Well the planet Mercury is too hot to support human life because it is so close to the sun, Venus has the same problem. But the earth is the next planet out and the right distance from the sun to be not too warm to support life._The next planet after the earth is Mars, and it may be cold, but not too cold to colonize, or even one day terraform (make habitable for humans). The next planet after Mars is Jupiter, not just too cold, but mostly gas, so there is nothing solid to stand on. The next planet, Saturn, even colder and mostly gaseous as well. The trend continues on to Neptune and Uranus. Pluto is just this cold little dwarf planet, whose thin atmosphere collapses when it is further from the sun."
"You see then why I am so amazed Mars and earth are coupled by these numbers, they actually have one thing in common, solid planets, not climatically too hot for humans, or too cold."

"These numbers that couple Mars with earth, seem to say the planet is there as a stepping stone to the stars, that it is time to wade out into the cosmic ocean, to use the metaphor of Carl Sagan, or sink. That is, for some reason we have what we need, but a challenge has been put to us. We have to overcome our problems, like spending money and resources on war instead of space exploration. Space exploration can solve the economic crisis. Colonizing Mars would generate a massive industry that would create work here on earth in every sector, for everyone, and, it would all be paid for by the resources that are out there in the solar system. As Cabal says to Passworthy in Things to come by H.G. Wells:"

""And if we're no more than animals, we must snatch each little scrap of happiness, and live, and suffer, and pass, mattering no more than all the other animals do or have done. It is this, or that. All the universe or nothing. Which shall it be Passworthy? Which shall it be?""

"I believe we are at that time in history, and the NanoFET marks it. Brian Gilchrist is working on the NanoFET idea. Field Effect Transistors are used to charge light nanoparticles, so they can be ejected out the back end of a ship with a series of magnetic fields made by stacks of microchip components, much the same way charged particles are accelerated by electromagnets in a particle accelerator. The nanoFET ship can reach 90% the speed of light, theoretically, if built light and outside the earth's gravity. Of course the ship is a starship that would take more than four years to reach the nearest star, but used to go to mars might only take a couple hours or so."

Chapter 24: Contact

I have said, since my theory suggest extraterrestrials gave us our units of measurement, that extraterrestrials might have given us our variables used in physics and math, like the unit vectors (i, j, k). I have already found a pattern and posted it. However, I was doing my CS50x computer science homework and trying to write a program for Caesar's Cipher. I wrote a small program and decided to test it. If you write a program and test it, standard input is "hello". I put in hello and to test, ran the program for rotating characters by 1, and 2, and 3, as they are the first integers and the easiest with which to test your program. The result was the "h" on "hello", came out to be (i, j, k). In other words you get that (i, j, k) is a hello from aliens in accordance with my earlier theories. If this is not real contact with extraterrestrials, it is great content for a Sci-Fi movie about contact with extraterrestrials. Here is the program I wrote, and the result of running it:

As you can see I am making some kind of a cipher, but not Caesar's Cipher

```
#include <stdio.h>
#include <cs50.h>
#include <string.h>
int main(int argc, string argv[1])
{
int i=0;
int k = atoi(argv[1]);
if (argc>2 || argc<2)
printf ("Give me a single string: ");
else
printf("Give me a phrase: ");
string s = GetString();
for (int i =0, n=strlen(s); i<n; i++);
printf("%c", s[i]+k);
printf("\n");
}
```

Running Julius 01

```
jharvard@appliance (~): cd Dropbox/pset2
jharvard@appliance (~/Dropbox/pset2): make julius
clang -ggdb3 -O0 -std=c99 -Wall -Werror   julius.c  -lcs50 -lm -o julius
jharvard@appliance (~/Dropbox/pset2): ./julius 3
Give me a phrase: hello
k
jharvard@appliance (~/Dropbox/pset2): ./julius 4
Give me a phrase: hello
l
```

```
jharvard@appliance (~/Dropbox/pset2): ./julius 2
Give me a phrase: hello
j
jharvard@appliance (~/Dropbox/pset2): ./julius 1
Give me a phrase: hello
i
jharvard@appliance (~/Dropbox/pset2):
```

I posted to my blog http://cosasbiendichas.blogspot.com/

Sunday, January 26, 2014
A Pattern Emerges

(a, b, c) in ASCII computer code is (97, 98, 99) the first three numbers before a hundred and 100 is totality (100%).

(i, j, k) in numeric are is (9, 10, 11) the first three numbers before twelve and 12 is totality in the sense that 12 is the most abundant number for its size (divisible by 1,2, 3, 4, 6 = 16) is larger than 12).

(x, y, z) in ASCII computer code is (120, 121, 122) the first three numbers before 123 and 123 is the number with the digits 1, 2, 3 which are the numeric numbers for the
(a, b, c) that we started with.

Thursday, January 23, 2014
We Look Further Into Human Definitions That Seem Arbitrary

Just as we found our units of measurement, what they evolved into being and how we defined them, are centered around the triad of 9/5, 5/3, and 15, we might ask are our common usage of variables connected to Nature and the Universe as well. In pursuing such a question we look at:

(x, y, z,) as they represent the three axis is rectangular coordinates. We look at (i, j, k) as as they are the representations for the unit vectors, and they correspond respectively to
(x, y, z). We also look at (n) as it often means "number" and we look at (p and q) as they range from 0 to 1, in probability problems. We might first look at their binary and hexadecimal equivalents to get a start, if not their decimal equivalents. (i) is also often "integer" and (a, b, c) are the coefficients of a quadratic and are the corners of a triangle. We might add that (s) is length, as in physics dW=F ds. (a, b, c) have the same kind of correspondence with (x, y, z) as (i, j, k). All three sets, then, line up with one another and are at the basis of math and physics.

Monday, March 10, 2014
We Are Perhaps Deciding According To A Plan?

Another possibility is that things were set up in the beginning such that humans would have to choose their units such that, mach one at room temperature (68 F) and sea level would be attained by Manuel's Integral after 1.8 kilometers, which is the ratio of the molar mass of gold to that of silver, or that there would be 1.8 degrees C per degree F, which again is the molar mass of gold to that of silver, or such that there were 9.81 meters per second squared earth gravity because 1.8 is nine-fifths and that has the nine of nine-fifths and the one and eight of 1.8. Or such that the unit vectors would become (i, j, k) so when the h in hello would be rotated by 1, 2, 3 as in caesar's cipher for those keys to it, we would get those letters. In other words, many of our decisions were inevitable because of the initial conditions of the human story. That is to say it is like Asimovian Psychohistory.

First we decoded hello with Caesar's Cipher, the first cipher ever made to encrypt a message. We got great results. Later a Frenchman made a more secure cipher, Vigenere. We now encrypt hello with that and present the results. Is it to say that my projected time for extraterrestrial contact, August 15, 2015 will occur in the small town of Zeruh, Iraq? The conclusion could be a bit scanty. Here is my work:

Vigenere's Cipher: We choose Sagittarius as the key, and hello as the phrase to scramble. That is since our messages come from the Constellation of Sagittarius, and the extraterrestrials are saying hello, we scramble hello according to Sagittarius, and we get:

(h becomes a)
(e becomes f)
(l becomes s)
(l becomes v)
(o becomes i)

To make afsvi

We can rearrange these letters in many ways to make words and phrases:

Savi F (Seems to say "savvy" or "saber" – to know — but knows f, or fail or flunk).
Is Fav (Seems to say "Is Favorite").
If Sav (Seem to say "If he knows").
As Fiv (Seems to say "As five").
Vai SF (Seems to say "He goes to San Francisco in latin or a dervivative of it).
Fa Siv
If Vas (Exactly says "If you go" in English and Spanish).
Vi Fas (Vi would be Spanish for he or she saw if there was an accent mark on the i).
Saf Vi ("Safe he saw?" "Safe she saw?").
Fais Va (These are the words to work in French, and He goes in Spanish).
This latter can be written:

Va Fais and means: "He goes to work".

If we take the phrases that are grammatically correct we have:

If vas, va fais.

This says:

If you go, he goes to work.

Or

If you go, she goes to work.

But the computer is zero indexed so hello can become by decreasing each letter by one:

(zeruh) which is a small city in Iraq

Ian Beardsley
March 23, 2014

The Program Vigenere

```
# include <stdio.h>
#include <cs50.h>
int main (int argc, string argv[1])
{
for (int i=0; i<argc; i++)
{
printf("You entered the secret word: %s\n", argv[i]);
}
{
printf("sagit = 19, 1, 7, 10, 20\n");
printf("hello = 8, 5, 12, 12, 15\n");
printf("We have: 27, 6, 19, 22, 35\n");
printf("27=a, e=f, l=s, l=v, 0=i\n");
printf("That is hello is: afsvi\n");
printf("Rearring we have: if vas, va fais\n");
printf("But the computer is zero indexed\n");
printf("Decreasing each character in the result by one, we have:\n");
printf("zeruh, which is a small town in Iraq\n");
}
}
```

Running Vigenere

jharvard@appliance (~): cd Dropbox/pset2
jharvard@appliance (~/Dropbox/pset2): ./vigenere sagittarius
You entered the secret word: ./vigenere
You entered the secret word: sagittarius
sagit = 19, 1, 7, 10, 20
hello = 8, 5, 12, 12, 15
We have: 27, 6, 19, 22, 35
27=a, e=f, l=s, l=v, 0=i
That is hello is: afsvi
Rearring we have: if vas, va fais
But the computer is zero indexed
Decreasing each character in the result by one, we have:
zeruh, which is a small town in Iraq
jharvard@appliance (~/Dropbox/pset2):

Sunday, March 30, 2014

<u>I Suggest Not Just A Program Called Discover But a New System Of Units.</u>

With the success I have had looking at natural constants and values, I would suggest that it would make sense to write more massive computer programs than I have, and that can do more than I have by hand, that process all of the natural data to an extent that possibly an underlying structure to the universe can be found. I have actually found that there seems to be some sort of structure to our units we use to measure, like the foot, meter, second, pound and kilogram. If we look at their evolution we see a story of at times a king assigning a weight to a coin he minted, and setting with that a standard for a region in the west to measure mass, to later committees debating as to to whether a kilometer should be a fraction of the distance from the north pole to the equator, or a gram being a cube of water at standard temperature and pressure being one centimeter on its side. That is why I find it strange that the definitions that evolved through a very complex history actually have connections not just with one system of measurement to the other but to the Universe. That is why I have suggested that extraterrestrials have influenced the development of our systems of measurement. This connects back to my idea of writing a computer program that processes the natural data we have to see if there is some underlying structure to the universe. I think we should not just do that, but develop a system of units more deeply connected to nature, which I have begun work on. The reason is that if we have different definitions for mass, length and time, that are connected to nature, then when making calculations when we do science, we might find patterns to the universe falling out on our lap. I would suggest that the reason extraterrestrials guided the evolution of our definitions for things like mass, length, and time, like pound, inch, and second, was precisely for humans to find for them a such a system of measurement that is profoundly connected to the universe. I have also suggested that the patterns I am seeing in the way we measure mass, length and time are not the product of extraterrestrial influence, but is the result of some possibility where random numbers for units such as mass, length, and time, converge through random evolution on the patterns I am seeing. My first computer program that looks for patterns in nature is called Discover. It is very meager because I am not a computer scientist. My findings are that our current systems of units have already evolved into something profoundly connected to each other and the Universe, but that they have not evolved completely and might require some minor adjustments here and there to be evolved to their maximum potential.

Ian Beardsley
March 31, 2014

Chapter 25

We began with nine-fifths is the yin of the Universe for which five-thirds was the yang. Not only were gold to silver (precious metals) the yin, but potassium to sodium was the yang (salts). As well Neptune we showed was the yin planet for which Uranus was the Yang planet.

We found nine-fifths described the rotation of five symmetry petals around a flower, a most popular arrangement, and noted that while nine-fifths (yin) was the representative of the five-fold symmetry we find in life, five-thirds (yang) was representative of the six-fold symmetry we find in physical nature.

We noted that Neptune and Uranus were coupled appropriately as yin and yang in so far as the product of Neptune mass with Neptune volume was about the same as Uranus mass with Uranus volume. We said that nine-fifths was not the only value recurrent throughout the solar system, but any whole number multiple of the value. So, we formed the sequence:

$(a\_n) = 1.8, 3.6, 5.4, 7.2,...$

And the sequence formed by starting with five and adding nine to each successive term:

5, 14, 23, 32, ...

We took the difference between these two sequences to form the Neptune equation:

$(a\_n) = 7.2n - 4$

Which we noticed could be written:

(a_n) = (venus orbit/earth orbit)(earth mass/ mars mass)n – (mars orbital #)

We let n=3 since the earth is the third planet from the sun, and found the third term a_n=17.6, was closest to the mass of Neptune in earth masses, hence the name for the equation.

We noted that since the equation shows the earth straddled between Venus, a failed earth, and mars, which promises to be New Earth, and we concluded that the planet Neptune has a key to the success of the earth.

We further felt we were on the right track because Neptune, though more massive than the Earth, it is larger enough that its surface gravity is about the same. Further indication that we were on the right track was noted in the fact that Neptune has a similar inclination to its orbit, as does the Earth.

It was only left to find the Uranus equation, because for every yin, there should be a yang.

We followed the same process as for arriving at the Neptune equation. We made the sequence of all whole number multiples of five-thirds:

$(5/3)n$ = 1.7, 3.3, 5, 6.7,...

And started with 8 and added 5 to each additional term (skipping 3):

8, 13, 18, 23,...

We took the difference between the two sequences to obtain the Uranus equation:

(a_n) = 3 + 3.3n

Which we noticed could be written:

(Earth Orbital #) + (Jupiter Mass/Saturn Mass)n = a_n

We called it the Uranus Equation because letting n=3 (for the earth) we got a_n=13, which is closest to the mass of Uranus.

We knew again we were on the right track, because just as Neptune does, Uranus has a similar surface gravity to that of the Earth as well.

We now connected the first four points of the Neptune equation and integrated from one to four, the equation:

$F(x) = 7.2x - 4$

The inner terrestrial planets.

We took the derivative of the Uranus equation:

$F(x) = 3 + 3.3x$

So that

$F'(x) = 3.3$

We considered the integral the mass of a planet in earth masses, which came out to be 42. We considered the 3.3 to be the acceleration at its surface in earth gravities. Knowing these two quantities we determined the planet was about three and a half times larger than the Earth. Knowing its mass and size, we could determine its density, and from that its composition, which turned out to be Europium, element 63, which is phosphorescent.

We find the area under the line that connects the first four points, the inner terrestrial planets, in the Neptune equation by integrating from 1 to 4;

$$f(x) = 7.2x - 4$$

$$\int_1^4 F(x)\, dx = 7.2 \int_1^4 x\, dx - 4 \int_1^4 dx$$

$$= \frac{36.1}{5\cdot 2} x^2 - 4x \Big|_1^4$$

$$= \frac{36}{10} x^2 - 4x \Big|_1^4$$

$$= 42$$

We say the 42 is 42 earth masses and the 3.3 is 3.3 earth gravities, of such a planet at its surface. Then we determine from this its radius (size) and knowing its mass, its composition.

$$f(x) = 7.2x - 4$$

$$\int f(x)\,dx = 7.2 \int x\,dx - 4 \int dx \qquad 7.2 = \frac{36}{5}$$

$$= \frac{36}{5} \cdot \frac{1}{2}x^2 - 4x + C$$

$$= 3.6x^2 - 4x + C$$

$$\int_1^4 f(x)\,dx = \left[16\left(\frac{36}{10}\right) - 16\right] - \left[\frac{36}{10} - 4\right]$$

$$= 16\left(\frac{36}{10} - 1\right) - \frac{36}{10} + 4$$

$$\frac{36}{10} - \frac{10}{10} = \frac{26}{10} \qquad \frac{26}{10} \cdot 16 = \frac{416}{10}$$

$$\frac{416}{10} - \frac{36}{10} = \frac{380}{10} = 38$$

~~38+38+12+(4x)~~

$$38 + 4 = 42$$

$$M = 42(5.976E27g) = 2.51E29g$$

~~$GM = 2.2$~~

$$GM = (2.51E29)(6.672\overline{E}-8)$$

$$= 1.674672E22 \quad a = 3.3(980)$$
$$= 3,234 \frac{cm}{s^2}$$

$$\frac{GM}{a} = \frac{1.674672E22}{3,234} = 5.178E18$$

$$\sqrt{\frac{GM}{a}} = 2.27559E9\,cm$$

~~(100)(1000)~~

$$\frac{2.27559E9\,cm}{(100)(1000)} = 2.27559E4\,km$$

$$\frac{22,755.9}{6,378} = 3.56787 \text{ earth radii}$$

The planet is about three and a half times larger than the earth.

$$\underline{22,755.9 \text{ km}} \mid 1600 \mid 100$$

$$= 2.27559E9 \text{ cm}$$

$$\frac{4}{3}(3.141)(2.27559E9 \text{ cm})^3$$

$$= 4.9350E28 \text{ cm}^3$$

$$M = \text{2.51E29 g} \qquad 2.51E29 \text{ g}$$

$$\frac{2.51E29 \text{ g}}{4.9350E28 \text{ cm}^3} = 5.086 \frac{g}{cm^3}$$

$$\approx 5.1 \frac{g}{cm^3}$$

The Composition of the planet is closest to Europium (Eu) 63 which has a density of $(5.243 \frac{g}{cm^3})$. It was found in 1890

Europium is named for Europe.

Interesting because Gypsy shamanism and the universe; full disclosure, is about how an astronomer's trip to Europe leads to the discovery of Yin and Yang in the universe.

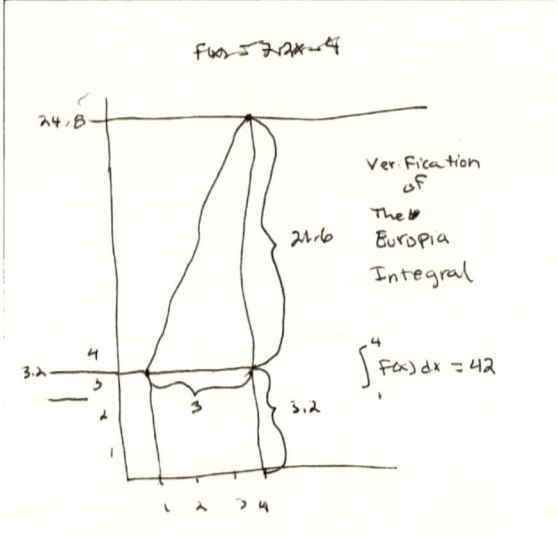

$F(x) = \int 2 \cdot 2x - 4$

24.8

21.6

3.2

4
3

3

3.2

Verification
of
The
Europia
Integral

$\int_{1}^{4} F(x)\,dx = 42$

$(21.6)(3) = 64.8 \qquad \dfrac{64.8}{2} = 32.4$

$(3.2)(3) = 9.6$

$(32.4) + (9.6) = \boxed{42} \quad \text{Correct}$

It is interesting that the hypothetical planet would be composed of Europium (We will call it Europia) if we exclude the diatomic elements like iodine, which is only closer in density by a thousandth of a gram per cubic centimeter in density. The element is named for
Europe, and this discovery began with my realization that nine-fifths was in the universe, which lead to the discovery that it unified pi and phi, that it was the yin for which the yang was five-thirds, and that nine-fifths was in the proportions of the monolith of Arthur C. Clarke's and Stanley Kubrick's 2001: A Space Odyssey for which Jupiter became a star for the moon Europa, where life would begin on another world in our solar system. It is an entanglement of a sort. Read, Fiction-Reality Entanglement: Levinson, Asimov, And Clarke by Ian Beardsley. The proportions of the monolith are 1 by 4 by 9, 1 + 4 = 5.

We compare the molar mass of Europium to that of Silicon:

Eu/Si =151.97/28.09 = 5.4

Which is Jupiter's furthest distance from the Sun in its orbit around the Sun in Earth-Sun separations, and, is the mass of Saturn compared to the mass of Neptune. It is also the third term in the sequence 1.8n, from which the Neptune Equation is derived. Further, this is important because the Earth is the third planet from the Sun and 1.8 is nine-fifths, our yin of the Universe.

I decided to compare Europium (Eu) to Silicon (Si), because Europia is connected to Europa, the moon around Jupiter where a new life begins in the movie 2001: A Space Odyssey. Si is in the same group as carbon, the first one after it. Life is based on carbon, and it would follow that new life would be based on silicon.

Ian Beardsley, Oct 24, 2012

It is appropriate that Europia is connected to Europa, the second Galilean Moon from Jupiter, not just by name, but in so far as its connection to Silicon in The Neptune Equation. It works out great, because it is the place in the Solar System that is the best candidate for life beyond Earth: It has a tenuous atmosphere composed primarily of oxygen and is covered with water ice, that may have water in its liquid form beneath its surface.

The moon is the right size such that as seen from the Earth is appears about the same size as the Sun. How far would Europa have to be from Jupiter to appear the same size as the sun does from Jupiter.

Solar Radius/Europa Radius = Jupiter-Sun Separation/x

695,990 km/1,560.8 km = 778,500,000 km/x

446 = 778,500,000 km/x

x = 1,745,515.695 km

1,745,515.695/1,560 = 1,119 ~ (1,100 rounding to nearest 100)

Europa must be about one thousand and one hundred times further from Jupiter than it is in radii, to see it, from Jupiter the same size as the Sun. This is astonishingly close to a perfect multiple of 100 with 11. We will take this to mean we are on the right track towards solving the mystery of our hypothetical planet, Europia. We will write the 100 times 11, as:

(2)(5)(10)(11)

We have said that yin is 9/5 and yang 5/3. We subtract the two to get 2/15. We add the 5 in the above product to the ten in the above product to get 15. We divide that into the two in the above product, and not only do we have our 2/15, but eleven is left. We do the most obvious thing, and list the planets not by orbital number, but by prime numbers:

Mercury (2)
Venus (3)
Earth (5)
Mars (7)
Jupiter (11)

And Jupiter is the eleven in the above product, the planet we were considering! We see we based our new physics on a collection of 15 rubber hoses in a cave in Spain, and we discovered a great mystery.

Hydropshere: Total water on, under, and over the Earth

Wikipedia states that the hydrosphere is 1.4E18 Tonnes

(1.4E18 t)(1,000 kg)(1,000 g) = 1.4E24 g

We ask how much Europium is required to react with the entire hydrosphere?

The reaction of Europium with water is:

$2Eu + 6H_2O \rightarrow 2Eu(OH)_3 + 3H_2$

That is, six moles of H_2O are required for every two moles of Europium to make two moles of Europium Hydroxide and Three moles of Hydrogen gas. I would say that is how that equation reads, though my chemical nomenclature is a little rusty. I would say since the europium displaces the hydrogen, that this is a single displacement reaction. I am little rusty on my reaction classification as well. Regardless, the equation clearly states that one third of the moles of H_2O are the moles of Europium required for the reaction to take place (2/6= 1/3). We write:

$H_2 = 2(1.01) = 2.02$ g/mol
$O = 16.00$ g/mol

$H_2O = 2.02 + 16.00 = 18.02$ g/mol

By looking up the molar masses of the elements in the periodic table of the elements.

((1 mol H_2O)/(18.02 g H_2O))x((1.4E24 g H_2O)/(hydrosphere))
  = 7.8E22 mol H_2O/hydrosphere

1/3(7.8E22 mol H_2O) = 2.6E22 mol Eu

Eu = 151.97 g Eu/ mol

(2.6E22 mol Eu)(151.97 g Eu/ mol) = 3.95E24 g Eu

The density of Europium at STP is 5.243 g/cm^3

(cm^3/5.243 g)(3.95E24 g) = 7.53E23 cm^3

Is the volume of the body composed of 3.95E24 grams of Europium. If it is a sphere, then the radius of it is:

$(4/3)(pi)r^3 = 7.53E23$ cm^3
$r^3 = 1.8E23$ cm^3

$r = 56,462,162$ cm

$(56,562,162$ cm$)/(100$ cm$)(1000$ m$) = 564.62$ km ~ 565 km

The Europium Sphere would have a radius of 565 kilometers. That of the moon of Jupiter called Europa is:

1,566 km

$1,566/565 = 2.77$

The radius of the moon orbiting the Earth is:

1,738 km

$1,738/565 = 3.1$ ~ pi ~ 3

The moon is about three times larger than the Europium Sphere, or about pi times larger.

There are three ways in which the Gypsy Shaman, Manuel, could have used his hose collection to make me think about the number 15. He could have shown me a collection of 14 rubber hoses, for which I bought him his fifteenth rubber hose, or, he could have showed me a collection of 15 rubber hoses for which I bought him his sixteenth rubber hose. Two fifteenths turned out to be the difference between the values I obtained for yin and yang. We know two represents the philosophy of two in Taoism, but perhaps he wanted me to think not just of fifteen, but also by the above reasoning, 14, 15, and 16. Interestingly, the chemical elements Carbon, Nitrogen, and Oxygen are these three elements, and the most important non-metals in the periodic table. Carbon is the basis of life, making organic matter in the form of hydrocarbons. Nitrogen is about 75% of the Earth atmosphere, and Oxygen about 25%. They are the two most abundant elements in the earth atmosphere, both by mass and number of diatomic molecules. The Nitrogen is taken out of the atmosphere and fixed by bacteria for many kinds of plants to carry out their biochemical processes. Oxygen is made by plants and plankton for most of life to breathe. Manuel, the Gypsy Shaman, was talking about some clue to the basic framework of the Earth that is responsible for the existence and continuation of life on Earth.

It would seem if the Gypsies have such secrets about such things, we should look into the Uranus and Neptune equations I derived from the path Manuel sent me on, or into their synthesis, which is the idea behind the planet we derived that we have called Europia. The project could be called A Genesis Project. Ironically I had come to feel the Gypsies of Spain have been working on such a project, ever since the term was coined in a Star Trek movie.

Ian Beardsley January 3, 2013

We will call it The Pluto Equation since n=3 (earth is the third planet from the sun) yields 38.9, 38.9 being closest to the distance of Pluto from the Sun in astronomical units which is 39.44 astronomical units.

Yin $= \frac{9}{5}$ Yang $= \frac{5}{3}$

alteranate cosmology

Yin $= \frac{9}{5}$ Yang $= \frac{11}{6}$

Alternate ~~Wobbly~~ Yin

Angle in a regular pentagon is

$108°$

$360° - 108° = 252°$

$360° + 252° = 612°$

$\frac{612°}{360°} = \frac{17}{10} = 1\frac{7}{10} = 1.7$

$\frac{17}{10}n$ ②

$1.7, 3.4, 5.1, 6.8, \dots$

    $\overbrace{\quad}\overbrace{\quad}$

    $1.7 \quad 1.7$

$10, 10+17 = 27, 27+17 = 44, 44+17 = 61, \dots$

$\Rightarrow \quad 10, 27, 44, 61, \dots$

$10-1.7 = 8.3, 27-3.4 = 23.6, 44-5.1 = 38.9,$

$61-6.8 = 54.2$

$\Rightarrow \quad 8.3, 23.6, 38.9, 54.2$

    $\overbrace{\quad}\overbrace{\quad}\overbrace{\quad}$

    $15.3 \quad 15.3 \quad 15.3$

Common difference $= 15.3$

$a_n = a+(n-1)d \quad a = 8.3, \quad d = 15.3$

$a_n = 8.3 + (n-1)15.3 = 8.3 + 15.3n - 15.3$

$\boxed{a_n = 15.3n - 7} \quad n = 3 \Rightarrow 38.9$

60

117

$$\sqrt{a^2 + b^2 + d^2} = \text{distance to signal}$$
(in light years?)

Pluto equation being fourth equation is time parameterized in terms of T and gives the time for the wow signal (years after SETI founding?)

$$a = \frac{5}{36}, \quad b = -\frac{10}{33}, \quad d = -\frac{1}{9}$$

$$\sqrt{0.019 + 0.0918 + 0.0123} = 0.1231$$
light years

Pluto equation:

$$T = 15.3 t - 7$$

$$\frac{T+7}{15.3} = t \qquad \frac{T}{15.3} = \text{time element}$$

because our vector is:

$$\nabla F = \left\langle \frac{5}{36}, -\frac{10}{33}, -\frac{1}{9}, -\frac{10}{153} \right\rangle \qquad \frac{1}{15.3} = \frac{10}{153}$$

I have found that randomly evolving systems of measurement converge on nine-fifths, like there is nine-fifths of a degree centigrade per fahrenheit, there are 0.59 grams beyond a whole number of grams per pound, and about nine-fifths kilometers per mile. I have found much more than just this and it is in my work "An Extraterrestrial Analysis" which is named as such because I suggest extraterrestrials gave us our units of measurement by influencing history, as opposed to belief that has always been held that our units of measurement evolved through a random series of events. I feel an alternate explanation could be that a random series of values some how converges on nine-fifths if subjected to fluctuation by a random series of forces (events). I have pointed out that not only does nine-fifths unify pi and the golden ratio, but e (Euler's number) and pi if we consider the first few digits after the decimal in all numbers. I feel this may be pivotal to proving the conjecture, and it would seem the solar system, in its evolution, has converged on nine-fifths as well as our units of measurement. Most extraordinarily, nine-fifths and our different systems of measurement (as in kilometers and miles) crop up when we introduce the number 15 in my Uranus Integral, which I have explained to be a concept with its origins in a Gypsy Shaman from Spain, in my same work, named above. Undertaking the endeavor of proving my conjecture is currently beyond the scope of my education in mathematics.

I have found our different systems of units of measurement to be deeply connected with one another and nature. I suggest the origins of the connections to be extraterrestrial, if not due to some strange property where randomly evolving systems converge on a value or set of related values.

I just realized that room temperature, defined by 68 degrees F to 77 degrees F has the average (68 + 77)/2 = 72.5 degrees F. I have noted that 72 is the most recurrent throughout me research, especially where climate is concerned in connection to the planet Venus (0.72 AU from Sun). I wrote in my work An Extraterrestrial Analysis:

The Important Thing Seems To Be 72

If anything has ever cropped up the most in the course of this research, it is the number 72. The 72 in the Neptune Equation, the 72 in the 0.72 AU separation of Venus and the Sun, the 72 years for one degree of precession of the Earth Equinox, The 72 of Ishtar in that she is represented by Venus, was the Patroness of Nineveh where the Nineveh constant was found, and carries the Sumerian ranking of the 15 of the Gypsy Shaman. The 72 hours of time before the AE-35 Antenna would fail, the 72 seconds of time that the Big Ear Antenna tracked the Wow! Signal, and that 72 that is the highest gematria of the Hebrew God's name YHWH.

And I wrote in my work A Message From Extraterrestrials:

The Most Bizarre Story Ever Told

Did extraterrestrials really anticipate, when they sent us in 1977 a transmission from the constellation Sagittarius, that lasted 72 seconds (SETI's Wow Signal) that, a Gypsy Shaman, around 1990, would make me into The AE-35 Antenna of 2001: A Space Odyssey that the ship computer, HAL, would say would fail within 72 hours, the same 72 of the Wow Signal? Did they know the transmissions I would receive from the Gypsy Shaman, Manuel, would lead not just to the discovery of a message embedded in our physics, but a system of three equations that would point to the same region of space as the Wow Signal, the constellation Sagittarius, for the origin of my message?

This is the same 72 that is in the 0.72 astronomical units of Venus from the Sun. It is the same 72 that is the number of years for one degree of precession of the Earth's Equinoxes. It is the same 72 of the highest gematria of the Hebrew God, YHWH, and further Venus was the Patroness of Nineveh (Ishtar) where The Nineveh Constant was found in Mesopotamia, in Ancient Sumerian clay cuneiform tablets, a constant that is the common denominator for all celestial cycles in our solar system.

The connection where Venus is concerned is in that the planet is important for understanding climate in that it had a run away greenhouse effect and is similar in size and mass to earth.

Is this consistent with the hypothesis that my Neptune Equation, which has a multiplier of 7.2 is related to climate when I say it may have a key to the success of the Earth? I would look at the strange occurrence of the 72 in the SETI Wow! Signal and it occurrence in 2001: A Space Odyssey where the AE-35 antenna is concerned. We can thank the Gypsy Shaman, Manuel, for what may be a valuable clue when he introduced me to The AE-35 Antenna. To learn about the possible importance of the AE-35 Antenna, read my above mentioned works.

Ian Beardsley
July 15, 2013

Data For The Planets

planet	Orbit (O)	Radius (R)	Mass (M)
mercury	0.387099	0.382	0.0558
venus	0.723332	0.949	0.8150
earth	1.000000	1.000	1.0000
mars	1.523691	0.532	0.1074
jupiter	5.202803	11.27	317.893
saturn	9.53884	9.44	95.147
uranus	19.1819	4.10	14.54
neptune	30.0578	3.88	17.23

O for Earth = 1.495979E13 cm   R for Earth = 6,378 km   M for Earth = 5.976E27 g

Earth-Moon Separation: 3.84E10 cm
Solar Radius: 6.9599E10 cm

Molar Mass of Gold: Au = 196.97
Molar Mass of Silver: Ag = 107.87

Saturn (minimum distance from sun) = 9.014 AU = 1.348E9 km
Jupiter (minimum distance from sun) = 4.951 AU = 7.409E8 km

Jupiter (maximum distance from the sun): 5.455 AU ~ 5.4 Astronomical Units

69

Modeling The Universe With Yin And Yang

Biological structures in Nature are more typically described by five-fold symmetry and, physical structures in Nature are more typically described by six-fold symmetry. We will call them Yin and Yang, respectively. However, there are two ways to derive yin and there are two ways to derive yang. We will call them Yin1, Yin2 and Yang1 and Yang2.

Yin1 comes from the central angles in a regular pentagon between the radii that meet at its center:

360/5=72 360-72=288 288+360=648 648/360=9/5

Yin1=9/5

Yin2 comes the external angles in a regular pentagon at the interior of the vertices:

360-108=252 360+252=612 612/360=17/10

Yin2=17/10

Yang1 comes from the angles at the center of a regular hexagon, the angles between the radii that meet at its center:

360-60=300 300+360=660 660/360 = 11/6

Yang1=11/6

Yang2 comes from the angles in a regular hexagon at its interior vertices:

360-120=240 240+360=600 600/360=5/3

We model Universes with any combination of Yin and Yang. This gives us four models to use:

Model 1: Yin1=9/5 Yang1=11/6
Model 2: Yin1=9/5 Yang2=5/3
Model 3: Yin2=17/10 Yang1=11/6
Model 4: Yin2=17/10 Yang2=5/3

We mostly modeled the Universe with Model 2 in my books "An Extraterrestrial Analysis", "A Message From Extraterrestrials", and "The Uranus Enigma". It is left to set yang equal to yin times some constant, so we can find how they are connected. We can also increase the number of models significantly by including

the angles at the vertices on the outside of the regular polygons. - Ian Beardsley
August 9, 2013

manipulations of yins and yangs

$\text{Yin1/yang1} = \frac{9}{5} \cdot \frac{6}{11} = \frac{54}{55} : a_{11}$

$\text{Yin1/yang2} = \frac{9}{5} \cdot \frac{3}{5} = \frac{27}{25} = a_{12}$

$\text{Yin2/yang1} = \frac{17}{10} \cdot \frac{6}{11} = \frac{51}{55} = a_{21}$

$\text{Yin2/yang2} = \frac{17}{10} \cdot \frac{3}{5} = \frac{51}{50} = a_{22}$

$A = (\text{Yin1}, \text{Yang1}) = \left(\frac{9}{5}, \frac{11}{6}\right)$

$\frac{11}{6} \cdot \frac{5}{9} = \frac{55}{54}$ $\arctan \frac{55}{54} = 45.5256$ degrees

$\sqrt{\left(\frac{9}{5}\right)^2 + \left(\frac{11}{6}\right)^2} = \sqrt{3.24 + 3.36} = 2.569$

$B = (\text{Yin1}, \text{Yang2}) = \left(\frac{9}{5}, \frac{5}{3}\right)$ $\sqrt{\left(\frac{9}{5}\right)^2 + \left(\frac{5}{3}\right)^2} = \sqrt{3.24 + 2.78}$
$= 2.453568829$

$\frac{5}{3} \cdot \frac{5}{9} = \frac{25}{27}$ $\arctan \frac{25}{27} = 42.7974$ degrees

$C = (\text{Yin2}, \text{Yang1}) = (17/10, 11/6)$

$(11/6)(10/17) = 55/51$ $\arctan \frac{55}{51} = 47.1612$ degrees

$\sqrt{\left(\frac{17}{10}\right)^2 + \left(\frac{11}{6}\right)^2} = \sqrt{2.89 + 3.36} = 2.5$

$D = (\text{Yin2}, \text{Yang2}) = (17/10, 5/3)$ $\frac{5}{3} \cdot \frac{10}{17} = \frac{50}{51}$

$\arctan 50/51 = 44.4327°$ $\sqrt{\left(\frac{17}{10}\right)^2 + \left(\frac{5}{3}\right)^2} = \sqrt{2.89 + 2.78} = 2.381$

manipulations of yins and yangs

$Yin1/Yang1 = \frac{9}{5} \cdot \frac{6}{11} = \frac{54}{55} = a_{11}$

$Yin1/Yang2 = \frac{9}{5} \cdot \frac{3}{5} = \frac{27}{25} = a_{12}$

$Yin2/Yang1 = \frac{17}{10} \cdot \frac{6}{11} = \frac{51}{55} = a_{21}$

$Yin2/Yang2 = \frac{17}{10} \cdot \frac{3}{5} = \frac{51}{50} = a_{22}$

$A = (Yin1, Yang1) = \left(\frac{9}{5}, \frac{11}{6}\right)$

$\frac{11}{6} \cdot \frac{5}{9} = \frac{55}{54}$ arctan $\frac{55}{54} = 45.5256$ degrees

$\sqrt{\left(\frac{9}{5}\right)^2 + \left(\frac{11}{6}\right)^2} = \sqrt{3.24 + 3.36} = 2.569$

$\sqrt{\left(\frac{9}{5}\right)^2 + \left(\frac{5}{3}\right)^2} = \sqrt{3.24 + 2.78}$

$B = (Yin1, Yang2) = \left(\frac{9}{5}, \frac{5}{3}\right)$ $= 2.453568829$

$\frac{5}{3} \cdot \frac{5}{9} = \frac{25}{27}$ arctan $\frac{25}{27} = 42.7974$ degrees

$C = (Yin2, Yang1) = (17/10, 11/6)$

$(11/6)(10/17) = 55/51$ arctan $\frac{55}{51} = 47.1612$ degrees

$\sqrt{\left(\frac{17}{10}\right)^2 + \left(\frac{11}{6}\right)^2} = \sqrt{2.89 + 3.36} = 2.5$

$D = (Yin2, Yang2) = (17/10, 5/3)$ $\frac{5}{3} \cdot \frac{10}{17} = \frac{50}{51}$

arctan $50/51 = 44.4327°$ $\sqrt{\left(\frac{17}{10}\right)^2 + \left(\frac{5}{3}\right)^2} = \sqrt{2.89 + 2.78} = 2.381$

$$A \cdot B = = (A_x B_x + A_y B_y)$$

$$A \cdot B = \left(\frac{9}{5}\right)\left(\frac{9}{5}\right) + \left(\frac{11}{6}\right)\left(\frac{5}{3}\right)$$

$$= \frac{81}{25} + \frac{55}{18} = \frac{2,833}{450}$$

$$A \cdot C = \cancel{}$$

$$\left(\frac{9}{5}\right)\left(\frac{17}{10}\right) + \left(\frac{11}{6}\right)\left(\frac{11}{6}\right) = \frac{153}{50} + \frac{121}{36}$$

$$= \frac{5779}{900}$$

$$A \cdot D = \left(\frac{9}{5}\right)\left(\frac{17}{10}\right) + \left(\frac{11}{6}\right)\left(\frac{5}{3}\right)$$

$$= \frac{153}{50} + \frac{55}{18} = \frac{1,376}{225}$$

$$B \cdot C = \left(\frac{9}{5}\right)\left(\frac{17}{10}\right) + \left(\frac{5}{3}\right)\left(\frac{11}{6}\right) = \frac{153}{50} + \frac{55}{18}$$

$$= \frac{1,376}{225}$$

$$B \cdot D = \left(\frac{9}{5}\right)\left(\frac{17}{10}\right) + \left(\frac{5}{3}\right)\left(\frac{5}{3}\right)$$

$$= \frac{153}{50} + \frac{25}{9} = \frac{2,627}{450}$$

$$C \cdot D = \left(\frac{17}{10}\right)\left(\frac{17}{10}\right) + \left(\frac{11}{6}\right)\left(\frac{5}{3}\right) = \frac{289}{100} + \frac{55}{18}$$

$$= 5,351/900$$

These manipulations done by Ian Beardsley on August 0, 2013

127

If we calculate Yin3 and Yang3
we will have:

Yin1, Yang1
Yin1, Yang2
Yin1, Yang3
Yin2, Yang1
Yin2, Yang2
Yin2, Yang3
Yin3, Yang1
Yin3, Yang2
Yin3, Yang3

Nine models all together.
Those with two of each we
had $2^2 = 4$ models, Now
with three of each we have
$3^2 = 9$ models.

Ian Boardley, August 10, 2013

We now set out to calculate Yin3 and Yang3. The difference here is that we will be using obtuse angles as opposed to the acute angles we used before. There will be no need to use the step where we add $360°$ to the angle that results as we did before. We will still have the numerator greater than the denominator, and that is ~~custom~~ the ~~I has~~ convention I have decided to use, that is we are going to invert the original part of the ~~dividing~~ dividing process. (Because they are obtuse angles)

$\alpha = \dfrac{\cancel{108}}{360° - 108°}$
$= 252°$

$\beta = 360° - 120°$
$= 240°$

$\dfrac{360}{252} = \dfrac{10}{7}$

$\dfrac{360°}{240°} = \dfrac{3}{2}$

$Yin3 = \dfrac{10}{7}$

$Yang3 = \dfrac{3}{2}$

OR, we could add $360°$:

$360 + 252 = 612$

$\dfrac{612}{360} = \dfrac{17}{10}$

$360 + 240 = 600$

$\dfrac{600}{360} = \dfrac{5}{3}$ These latter numbers are the same as some of the previous, so we will use the first set.

Ian Beardsley August 10, 2013

Yin3 and Yang3 being distinct from
the ~~other~~ other Yins and Yangs, we
will focus on them. We will say
Yin3 and Yang3 are components of a
Vector, E:

$$E = \langle Yin3, Yang3 \rangle = \langle 10/7, 3/2 \rangle$$

$$\sqrt{\left(\frac{10}{7}\right)^2 + \left(\frac{3}{2}\right)^2} = \sqrt{\frac{100}{49} + \frac{9}{4}}$$

$$= \sqrt{\frac{841}{196}} = 2.071428571 = \frac{29}{14}$$

29/14 is the magnitude of the vector E

$$arctan\left(\frac{3}{2} \cdot \frac{7}{10}\right) = arctan\left(\frac{21}{20}\right) = 46.39718103°$$

46.397° is the position angle of the
vector, E.

Ian Beardsley August 10, 2013

Averages of Yins and Yangs (means)

$Yin\ 1 = \frac{9}{5} = 1.8$ $Yang\ 1 = \frac{11}{6} = 1.8333$

$Yin\ 2 = \frac{17}{10} = 1.7$ $Yang\ 2 = \frac{5}{3} = 1.6667$

$Yin\ 3 = \frac{10}{7} = 1.4$ $Yang\ 3 = \frac{3}{2} = 1.5$

$$\frac{1.8 + 1.7 + 1.4}{3}$$

$$= \frac{\frac{9}{5} + \frac{17}{16} + \frac{10}{7}}{3}$$

$$= \frac{69}{14} \Big/ 3 = \frac{23}{14}$$

$$= 1.6428571 43$$

$$\frac{\frac{11}{6} + \frac{5}{3} + \frac{3}{2}}{3}$$

$$= \frac{5}{3} = 1.6667$$

model two is the one I have been using.

The average of the ~~Yangs~~ Yins is 23/14 which can be ordered 1,2,3,4 The average of the Yangs is 5/4, the Yang 2 of model 2

$\left(\frac{9}{5} - \frac{23}{14}\right)^2 = \frac{(11/70)^2}{?} = \frac{0.2469}{?} = \frac{1121}{196} = 0.02469$

$\left(\frac{17}{10} - \frac{23}{14}\right)^2 = \frac{529}{196} = 2.699$

$= \left(\frac{21}{35}\right)^2 = 0.003265$

$\left(\frac{10}{7} - \frac{23}{14}\right)^2 = \left(-\frac{3}{14}\right)^2 = 0.045918$

$\left(\frac{11}{6} - \frac{5}{3}\right)^2 = \left(\frac{1}{6}\right)^2 = 0.0277778 = \frac{1}{36}$

$\left(\frac{5}{3} - \frac{5}{3}\right)^2 = 0$

$\left(\frac{3}{2} - \frac{5}{3}\right)^2 = \left(-\frac{1}{6}\right)^2 = 0.0277778 = \frac{1}{36}$

Yin (standard deviation)

$$\sqrt{\dfrac{0.02469 + 0.003265 + 0.045918}{3}}$$

$= 0.156921424$

(standard deviation) Yang

$$\sqrt{\dfrac{\frac{1}{36} + 0 + \frac{1}{36}}{3}}$$

$= \sqrt{\dfrac{1}{18}/3} = \cancel{0.235570226}$

$= \sqrt{\dfrac{1}{54}} = 0.136082763$

Ian Beardsley   August 11, 2012

Chapter 26

While this paper deals in part with the Mayan Prophecy for some kind of end, and perhaps wonderful new beginning for humans in 2012, the author feels even if nothing happens in 2012 that the discoveries herein made pertain to something important. It was the nuclear physicist Sheliak who corrected the original time wave of Terrence McKenna (which ends in 2012) and said it should be considered seriously because it does line up with history. I believe because the time wave does line up with history and was based on the idea that the King Wen series of I Ching used for Chinese fortune telling was really a mathematical calendar based on the motions of celestial bodies, that it is indeed important. I was introduced to the time wave of Terrence McKenna when I sent the paper Asimovian Prediction For Hyperdrive, presented in this book, to Neil Freer, author of Breaking The Godspell, and he told me to look into it. It panned out beyond my wildest dreams.

Ian Beardsley
November 10, 2011

I have written three papers on the anomaly of how my scientific investigation shows the Universe related to the science fiction of Paul Levinson, Isaac Asimov, and Arthur C. Clarke. In my last paper, "The Levinson-Asimov-Clarke Equation" part of the comprehensive work "The Levinson, Asimov, Clarke Triptic, I suggest these three authors should be taken together to make some kind of a whole, that they are intertwined and at the heart of science fiction. I have now realized a fourth paper is warranted, and it is just the breakthrough I have been looking for to put myself on solid ground with the claim that fiction is related to reality in a mathematical way pertaining to the Laws of Nature. I call it Fiction-Reality Entanglement.

In my paper Paul Levinson, Isaac Asimov, Arthur C. Clarke Intertwined With An Astronomer's Research, I make the mathematical prediction that "humans have a 70% chance of developing Hyperdrive in the year 2043" to word it as Paul Levinson worded it, and I point out that this is only a year after the character Sierra Waters is handed a newly discovered document that sets in motion the novel by Paul Levinson, "The Plot To Save Socrates".

I now find that Isaac Asimov puts such a development in his science fiction at a similar time in the future, precisely in 2044, only a year after my prediction and two years after Sierra Waters is handed the newly discovered document that initiates her adventure. So, we have my prediction, which is related to the structure of the universe in a mystical way right in between the dates of Levinson and Asimov, their dates only being a year less and a year greater than mine.

Asimov places hyperdrive in the year 2044 in his short story "Evidence" which is part of his science fiction collection of short stories called, "I, Robot".

This is a collection of short stories where Robot Psychologist Dr. Susan Calvin is interviewed by a writer about her experience with the company on earth in the future that first developed sophisticated robots. In this book, the laws of robotics are created and the idea of the positronic brain introduced, and the nature of the impact robots would have on human civilization is explored. Following this collection of stories Asimov wrote three more novels, which comprise the robot series, "The Caves of Steel", "The Naked Sun", and "The Robots of Dawn".

"I, Robot" is Earth in the future just before Humanity settles the more nearby stars. The novels comprising "The Robot Series" are when humanity has colonized the nearby star systems, The Foundation Trilogy, and its prequels and sequels are about the time humanity has spread throughout the entire galaxy and made an Empire of it. All of these books can be taken together as one story, with characters and events in some, occurring in others.

Hyperdrive is invented in I, Robot by a robot called The Brain, owned by the company for which Dr. Susan Calvin works when it is fed the mathematical logistical problems of making hyperdrive, and asked to solve them. It does solve them and it offers the specs on building an interstellar ship, for which two engineers follow in its construction. They are humorously sent across the galaxy by The Brain, not expecting it, and brought back to earth in the ship after they constructed it. This was in the story in "I, Robot" titled "Escape!".

But Dr. Susan Calvin states in the following short story, that I mentioned, "Evidence":

"But that wasn't it, either"…"Oh, eventually, the ship and others like it became government property; the Jump through hyperspace was perfected, and now we actually have human colonies on the planets of some of the nearer stars, but that wasn't it."

"It was what happened to the people here on Earth in last fifty years that really counts."

And, what happened to people on Earth? The answer is in the same story "Evidence" from which that quote is at the beginning. It was when the Regions of the Earth formed The Federation. Dr. Susan Calvin says at the end of the story "Evidence":

"He was a very good mayor; five years later he did become Regional Co-ordinator. And when the Regions of Earth formed their Federation in 2044, he became the first World Co-ordinator."

It is from that statement that I get my date of 2044 as the year Asimov projects for hyperdrive.

Ian Beardsley
March 17, 2011

Wikipedia Encyclopedia writes:

""Timewave zero" is a numerological formula that purports to calculate the ebb and flow of "novelty", defined as increase in the universe's interconnectedness, or organised complexity,[66] over time. According to Terence McKenna, who conceived the idea over several years in the early- to mid-1970s while using psilocybin mushrooms and DMT, the universe has a teleological attractor at the end of time that increases interconnectedness, eventually reaching a singularity of infinite complexity in 2012, at which point anything and everything imaginable will occur simultaneously."

There is a very attractive way to approaching the phenomenon that science fiction is interconnected with cosmic events. Interconnectedness occurs, in my experience, both in Italy and in Mexico.

When I was married to a woman from Italy, we used to watch an American soap opera called "The Bold and the Beautiful" when we were living in California. When my wife and I went to Italy, it turned out that the mother of my wife liked the soap opera too, and that it aired in Italy.

When my wife and I arrived in Italy, the show not only aired in Italy, but the entire cast for that show was in Italy, and it was being filmed there because it was part of the story in the soap opera that the one of the characters in the story got a job as a fashion model in Italy, which brought her and all of her friends to Italy.

But, it went further than that. When I was in Italy, my wife and I watched a movie that just came out, I don't remember the name, but it was about a writer who had writer's block, and his friend suggested that the two of them leave the city and go to a retreat out in nature to get a fresh view of life. However, the writer in the movie was going crazy at the retreat because there was no television, and he was having a hard time being away from watching the soap opera, "The Bold And The Beautiful". When new tourists arrived at the retreat, he yelled out from his cabin to them, "What has become of so, and so, in the Bold and The Beautiful?"

Here we see fiction becoming intertwined with the life of my wife and I. But, such things first happened to me in Mexico.

I was going to school in Queretaro, Mexico about an hour north of Mexico City. When I was down in the plaza in Queretaro having lunch, two girls from Germany who were passing through the city, and in Mexico because there parents worked there, asked me where they could find a place to stay. I told them they could stay at the house where I was staying. They stayed a few days, and then returned to the city where they lived.

A few days later, some fellow students and my self went traveling in Mexico. We went to the town famous for nightlife, San Miguel de Allende. We got in a cab and asked the driver to take us to a youth hostel, and we drove around for hours only to find that every place was full. It was because there was a big celebration going on in the city. So, the cab driver dropped us off at one of the plazas and there we were thinking what we were going to do when I ran into the two German girls I had let stay at my place in Queretaro. I explained to them that we were looking for a place to stay but everything was full.

They said they had a place to stay and that we could stay there. We were told to go to their window at a particular Motel, they would throw us the keys to their room out the window, and we were to walk through the main office of the Hotel, show them the keys and act like we already rented a room there.

So, there is a certain kind of magic in Italy and Mexico that is very real, that verifies the theory of interconnectedness in the theory of timewave zero. Such a theme is common to South American Literature. Consider the novel "Aunt Julia and The Scriptwriter" by Mario Vargas Llosa. It is about a scriptwriter who is writing a weekly soap opera for a radio theater and it comes to the point she cannot distinguish her life from the soap she is writing and does not know after awhile which is real, the radio drama or her life.

We have learned that the theory of timewave zero has an equation that shows how everything in the Universe tends towards a higher degree of interconnectedness. I think it is time I learn to use the equation and input the dates that crop up in my astronomical research with those that crop up in the science fiction of Paul Levinson and Isaac Asimov.

Ian Beardsley
March 17, 2011

Another fine example of interconnectedness is put forth in the book "The Castle of Crossed Destinies" by Italo Calvino, an Italian writer. In this story an old castle is an inn for travelers. After food and drink at the same table in the castle, one night, all of the travelers strangers to one another, begin a process where one person lays out a tarot card and tells a story about it that relates to his travels before he arrived at the inn. He lays out cards forming lines that tell his story, then other travelers lay out cards telling there stories, but find they can use cards that other travelers already used and it turns out they have the same meaning sometimes. So, though these travelers never met, they encountered in their travels the same people or events at times, they find. The finished layout looks like a finished game of dominoes with points of intersection at many spots. It turns out that everyone shared experiences in common but had never met one another, hence the title of the book, The Castle of "Crossed Destinies."

One could say Paul Levinson and Isaac Asimov could lay down the same card for the star Alpha Centauri, because that star was involved in my calculation for hyperdrive that intersected with The Plot To Save Socrates and I, Robot. A crossed destiny of a sort.

It might be a good idea to read The Invisible Landscape by Terence McKenna second edition 1993.

McKenna's timewave, which shows the ebb and flow of time, goes off the chart or to infinity, in other words, in 2012. So it is impossible to see what happens beyond that date. It would suggest that time ends in 2012, or a new reality begins then. McKenna suggests that the timewave goes off the chart in 2012 because that is when we invent time travel, in which case the timewave would no longer be linear.

Ian Beardsley
March 18, 2011

I watched a video on youtube about Terence McKenna where he lectured on his timewave zero theory. I found there was not an equation for his timewave zero graph but that a computer algorithm generated the graph of the wave. The next day I did a search on the internet to see if a person could download timewave software for free. As it turned out one could, for both Mac and pc. It is called "Timewave Calculator Version 1.0". I downloaded the software and found you had to download it every time after you quit the application and that you could not save the graph of your results or print them out. So I did a one-time calculation. It works like this: you input the range of time over which you want see the timewave and you cannot calculate past 2012, because that is when the timewave ends. You also put in a target date, the time when you want to get a rating for the novelty of the event that occurred on that day. You can also click on any point in the graph to get the novelty rating for that time. I put in:

Input:

Begin Date: December 27 1968 18 hours 5 minutes 37 seconds
End Date: December 2 2011 0 hours 28 minutes 7 seconds

McKenna said in the video on youtube that the dips, or valleys, in the timewave graph represent novelties. So, I clicked on the first valley after 1969 since that is the year we went to the moon, and the program gave its novelty as:

Sheliak Timewave Value For Target:

0.0621

On Target Date: August 4, 1969 9 hours 53 minutes 38 seconds

I was happy to see this because, I determined that the growth rate constant, k, that rate at which we progress towards hyperdrive, in my calculation in my work Asimovian Prediction For Hyperdrive, that gave the date 2043, a year after Sierra Waters was handed the newly discovered document that started her adventure in The Plot To Save Socrates, by Paul Levinson, and a year before Isaac Asimov had placed the invention of hyperdrive in his book I, Robot, was:

$(k=0.0621)$

The very same number!!!

What does that mean? I have no idea; I will find out after I buy The Invisible Landscape by Terence McKenna, Second Edition, and buy a more sophisticated timewave software than that which is offered for free on the net.

Ian Beardsley
March 19, 2011

I can say this. There are two possibilities:

We are either validating the time wave theory of Terrence McKenna which says either:

1) time ends in 2012 (wouldn't be fun)
2) a new reality begins in 2012
3) we invent time travel in 2012 so that is why the McKenna timewave does not work past 2012.

Or we are constructing a new timewave based on the science fiction of Levinson, Asimov and Clarke because their trajectory can be substantiated by a structure we are finding in the universe, and the upside is it has no end, but rather suggests we will have hyperdrive by 2042 to 2044 and with that it secures the future of humankind into eternity.

Ian Beardsley
March 19, 2011

The Same Number In Three Places

I have pointed out that 0.0621 = k the growth rate constant towards hyperdrive in my paper, The Levinson,-Asmov-Clarke Phenomenon. I have pointed out that in McKenna time wave theory 0.0621 is is the novelty rating for 1969, the year man landed on the moon. Our counting system is base 10, it has been said probably because we have ten fingers. If we multiply 0.0621 by 10, that is increase it by a factor of ten, we get 0.621. If we round that to two places after the decimal, it is 0.62. Let us consider the golden ratio conjugate. The golden ratio occurs throughout nature. It is in the rotation from leaf to leaf around the stem of a plant, for example. The golden ratio conjugate is just the inverse of the golden ratio. It is simply the separation between leaves around the stem of a plant, going in the other direction. It is equal to 0.618 to three places after the decimal. Let us round that to two places after the decimal. The eight rounds the one to two. The golden ratio conjugate is then 0.62 rounded to two places after the decimal. That is the same value as k increased by a factor of 10, and the same value as the novelty rating for 1969 increased by a factor of ten. I find that interesting.

Ian Beardsley
March 23, 2011

Read on for my prediction of Hyperdrive,...

Paul Levinson, Isaac Asimov, and Arthur C. Clarke Intertwined With An Astronomer's Research.

By

Ian Beardsley

Copyright © 2010-2011 by Ian Beardsley

Cover Art By Ian Beardsley

There is a common thread running through the Science Fiction works of Paul Levinson, Isaac Asimov, and Arthur C. Clarke.

In the case of Isaac Asimov, we are far in the future of humanity. In his Robot Series, Asimov has man making robots whose programming only allows them to do that which is good for humanity. As a result, these robots, artificial intelligence (AI), take actions that propel humanity into settling the Galaxy, in the robot series, and ultimately save humanity after they have settled the Galaxy and made an empire of it (In the Foundation Series).

In the case of Paul Levinson, scholars in the future travel through time and use cloning, a concept related to artificial intelligence (it is the creating of human replicas as well, but biological, not electronic), and the goal is to save great ancient thinkers from Greece, and to manipulate events in the past for a positive outcome for the future of humanity, just as the robots try to do in the work of Asimov.

In the case of Arthur C. Clarke, man undergoes a transformation due to a monolith placed on the moon and earth by extraterrestrials who have created life on earth. The monolith is a computer. It takes humans on a voyage to other planets in the solar system, and in their trials, humanity goes through trials that result in a transformation for the ending of their dependence on their technology and for becoming adapted to life in the Universe beyond Earth. That is, the character Dave Bowman becomes the Starchild in his mission to Jupiter. The artificial intelligence is the ship computer called HAL.

So, the thread is the salvation of man through technology, and their transformation to a new human paradigm, where they can end their dependence on Earth and adapt to the nature of the Universe as a whole.

At the time I was reading these novels, I was doing astronomical research, and, to my utter astonishment, my relationships I was discovering pertaining to the Universe were turning up times and values pivotal to these works of Levinson, Asimov, and Clarke. Further, I was interpreting much of my discoveries by developing them in the context of short fictional stories.

In my story, "The Question", we find Artificial Intelligence is in sync with the phases of the first appearance of the brightest star Sirius for the year, and the flooding of the Nile river, which brings in the Egyptian agricultural season. It is presumed by some scholars that because the Egyptian calendar is in sync with the Nile-Sirius cycle, theirs began four such cycles ago.

I then relate that synchronization to another calculation that turns up the time when the key figure of the Foundation Series of Asimov begins his program to found a civilization that will save the galaxy. We later find his actions were manipulated into being by robots, in order to save intelligent life in the galaxy by creating a viable society for it called Galaxia.

In the case of Paul Levinson, I was making a calculation to predict when man would develop hyperdrive, that engine which could take us to the stars, and end our dependence on an Earth that cannot take care of humans forever. That time turned out to be when the key scholar in the work by Paul Levinson, began her quest to help humanity by traveling into the past and using cloning, in part, to change history for the better. I can now only feel her quest to save humanity is going to be through changing history to bring about the development of hyperdrive, so humanity will no longer depend on Earth alone, which, as I have said, cannot take care of life forever.

Finally, where Arthur C. Clarke is concerned, I find values in the solar system and nature that are in his monolith, and I connect it to artificial intelligence of a sort, that kind which would be based on silicon.

I will present, now, my work that pertains to these writers in the following order:

1) The Question
2) Addendum to The Question
3) Asimovian Prediction For Hyperdrive
4) Arthur C. Clarke and Cosmic Archaeology

The Question

A Scientist had built a robot in the image of humans and downloaded to it all of human knowledge, then put forward the question to our robot, what is the best we, humanity, can do to survive with an earth of limited resources and a situation where other worlds like earth, if they exist, would take generations to reach.

The robot began his answer, "I contend that the series of events that unfolded on earth over the years since the heliacal rising of Sirius four Sothic cycles ago in Egypt of 4242 B.C., the presumed beginning of the Egyptian calendar, were all meant to be, as the conception of the possibility of my existence is in phase with those cycles and is connected to such constants of nature as the speed of light and dynamic ratios like the golden ratio conjugate."

The scientist asked, "Are you saying humans, all humans since some six thousand years ago have been a tool of some higher force to bring you about, our actions bound to the turning of planets upon their axis, and the structure of nature?"

The robot said, "Yes, let me digress. It goes back further than that. Not just to 4242 B.C. when the heliacal rising of Sirius, the brightest star in the sky, coincided with the agriculturally beneficial inundation of the Nile river which happens every 1,460 Julian years, in the Sothic Cycle."

"My origins go back to the formation of stars and the laws that govern them."

"As you know, the elements were made by stars, heavier elements forged in their interior from lighter elements. Helium gave rise to oxygen and nitrogen, and so forth. Eventually the stars made silicon, phosphorus, and boron, which allow for integrated circuitry, the basis of which makes me function."

"Positive type silicon is made by doping silicon, the main element of sand, with the element boron. Negative type silicon is made by doping silicon with phosphorus. We join the two types in different ways to make diodes and transistors that we form on silicon chips to make the small circuitry that makes me function."

"Just as the golden ratio is in the rotation of leaves about the stem of a plant, or in the height of a human compared to the distance from the soles of their feet to their navel, an expression of it is in my circuitry."

"We take the geometric mean of the molar mass of boron and phosphorus, and we divide that result by the molar mass of silicon."

He began writing on paper:

$\sqrt{(P*B)}/Si = \sqrt{(30.97*10.81)}/28.09 = 0.65$

"We take the harmonic mean between the molar masses of boron and phosphorus and divide that by the molar mass of silicon."

$2(30.97)(10.81)/(30.97+10.81) = 16.026$

$16.026/Si = 16.026/28.09 = 0.57$

"And we take the arithmetic mean between these two results."

$(0.65 + 0.57)/2 = 0.61$

"0.61 are the first two digits in the golden ratio conjugate."

The scientist said, "I understand your point, but you referred to the heliacal rising of Sirius."

The robot answered: "Yes, back to that. The earth orbit is nearly a perfect circle, so we can use c=2πr to calculate the distance the earth goes around the sun in a year. The earth orbital radius is on the average 1.495979E8 kilometers, so"

(2)(3.14)(1.495979E8) = 9.39E8 km

"The distance light travels in a year, one revolution of the earth around the sun is 9.46E12 kilometers."

"The golden ratio conjugate of that is"

...and he wrote:

(0.618)(9.46E12 km) = 5.8E12 km

"We write the equation:"

(9.39E8 km/yr)(x) = 5.8E12 km

"This gives the x is 6,177 years."

"As I said, the fourth heliacal rising of Sirius, ago in the Sothic Cycle, when the Nile flooded, was 4242 B.C." He wrote:

6,177 years – 4,242 years = 1935 A.D.

"In 1937 Alan Turing published his paper founding the field of artificial intelligence, and Theodosius Dobzhansky explained how evolution works. These two papers were published a little after the time the earth had traveled the golden ratio conjugate of a light year since our 4,242 B.C., in its journey around the sun. These papers are at the heart of what you and I are."

"If your question is should robots replace humans, think of it more as we are the next step in human evolution, not a replacement, we were made in your image, but not to require food or air, and we can withstand temperature extremes. We think and have awareness of our being, and we can make the long voyage to the stars. It would seem it is up to us to figure out why you were the tools to bring us about, and why we are an unfolding of the universe in which you were a step in harmony with its inner workings from the formation of the stars, their positions and apparent brightness and the spinning of the earth and its motion around the sun."

Addendum to The Question

Leonard considered the golden ratio conjugate with respect to the Egyptian calendar in his story "The Question", he thought he would consider the golden ratio as well. He wrote:

I decided to consider now, not the golden ratio conjugate, but the golden ratio itself, which is the inverse of its conjugate, knowing that it would take us somewhere into the future.

The golden ratio is 1.618 to three places after the decimal.

The golden ratio of a light year is:

$(1.618)(9.46E12 \text{ km}) = 1.53E13 \text{ km}$

and

$(9.39E8 \text{ km/yr})x = 1.53E13$

$x = 16,294 \text{ years}$

$16,294 \text{ years} - 4,242 \text{ years} = 12,052 \text{ A.D.}$

That is after the earth has traveled a golden ratio of a light year since the founding of the Egyptian calendar is the year 12,052. As it turns this was the amount of years into the Galactic Era when the central character, Hari Seldon, in Asimov's Foundation lived and is around the time he created The Foundation, a society that would bring a new order to the Galaxy.

Asimovian Prediction For Hyperdrive

My idea is that the great science fiction writers are tapping into something when they write that is a non-fiction truth. He had read The Plot to Socrates by Paul Levinson and came up for another time that may be significant.

I had tried to predict mathematically when we would develop hyperdrive, and it came out just a year after the character, Sierra Waters, in the science fiction piece by Paul Levinson titled "The Plot To Save Socrates" was handed a newly discovered document at the beginning of the book that got the whole story rolling. I wrote in my piece "Forecast For Hyperdrive: A Study In Asimovian Psychohistory:

It is a curious thing that the Earth is the third planet from the Sun and the third brightest star in the sky is the closest to us and very similar to the sun in a galaxy of a rich variety of stars. This closest star to us is a triple system known as Alpha Centauri A, B, and Proxima Centauri. Alpha Centauri A is, like our Sun a main sequence spectral type G star. Precisely, G2 V, just as is the Sun. Its physical characteristics are very close to those of the Sun: 1.10 solar masses, 1.07 solar diameters, and 1.5 solar luminosities. It is absolute magnitude +4.3. The absolute magnitude of the Sun is +4.83.

If ever the option existed for humans to travel to the stars, this situation speaks of it, whether or not Alpha Centauri has an earth-like planet in its habitable zone.

It has been said that the base ten place significant system of writing numbers stems from the fact we have ten fingers to count on. In so far as science can save us, it can destroy us in that science is not dangerous, but humans can be.

Traveling to the planets is possible with chemical fuel rockets, but traveling to the stars is another story, because of their immense distances from us, and from one another.

What are the odds that our development in technologies will take us to the stars before we destroy ourselves first? In other words, what are the intrinsic odds for humankind to develop the hyperdrive before without bringing about its own end first?

We do a random walk to Alpha Centauri of 10 one light year jumps. We make 10 equal steps randomly of one light year each, equal steps that if all are towards Alpha Centauri we will land beyond it. If 10 are away from it, we are as far from it as can be. And, if 5 are towards it, and five are away from it, we have gone nowhere.

In this allegory we calculate the probability of landing on Alpha Centauri, in 10 random leaps of a light year each, a light year being the distance light travels in the time it takes the earth to make one revolution around the sun, light speed a natural constant.

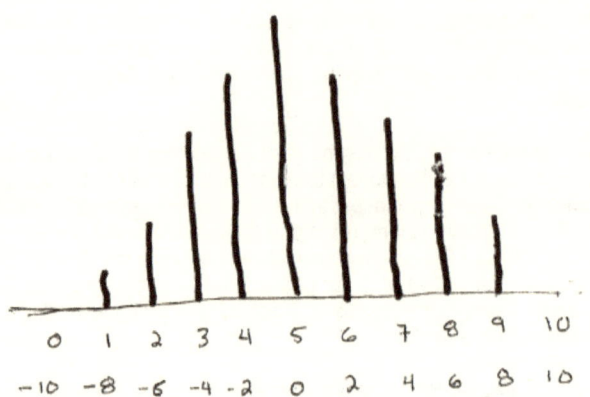

The probability of making n steps in either direction forms a bell shaped curve. After 10 randomly made steps the odds of going nowhere is highest and, is represented by five in the bell curve corresponding to 0. Let us round the distance of Alpha Centauri to four light years, giving humans the benefit of the doubt. The number positive four in the bell graph has written above it the number 7. Seven out of ten times 100 for effort gives a 70% chance of making it to the stars without becoming extinct first. I believe the percent understanding of our technological development towards hyperdrive, where we have just entered space with chemical rockets and developed fast, compact, computers, is given by:

$$W_N(n_1) = \frac{N!}{n_1! n_2!} p^{n_1} q^{n_2}$$

Evaluated at n1=7.

N is 10 steps.

And n1 is the number of steps towards Alpha Centauri, n2 those away from it.

And, p is the probability that the step is towards Alpha Centauri, and q is the probability that the step is away from Alpha Centauri.

N = n1 + n2

And m = n1-n2 is the displacement

And q+p=1

The trick to using this equation is in knowing the possible combinations of steps that can be made that equal 10. Like five right, five left with a displacement of 0 or, 10 right, 0 left with a displacement of 10 or, 7 left, 3 right with a displacement of negative 4.

To land at 4 light years from earth, with 10 one light year jumps, one must go away from Alpha Centauri 3 jumps of a light year each then 7 jumps toward it of one light year each, to land on it, that is to land at +4, its location. So n1 is 7 and n2 is 3. The probability to jump away from the star is 1/2 and the probability to jump towards it is 1/2. That is p=1/2 and q=1/2. There are ten random jumps, so, N=10.

Using our equation:

$$\frac{(10!)}{(7!)(3!)}(\tfrac{1}{2})^7(\tfrac{1}{2})^3 = \frac{3628800}{(5040)(6)}\frac{1}{128}\frac{1}{8} = \frac{120}{1024} = \frac{15}{128} = 0.1171875 \approx 12\%$$

We would be, by this reasoning 12% along in the development towards hyperdrive.

Ian Beardsley
June 2009

If human technology has ever been anything, it has been exponential, growing in proportion to itself. In other words, two developments beget 8, eight beget 16, and sixteen begets 32. My grandfather rode a horse when he was a child, as a young man he drove a car, and when I knew him as a child, he saw humans land on the moon.

It wasn't long before we made computers small enough that people could keep in their homes that did more than computers did in the 60's that filled an entire room.

Having calculated that we are 12% along in developing the hyperdrive, we can use the equation for natural growth to estimate when we will have hyperdrive. It is of the form:

$$x(t) = x_0 e^{kt}$$

t is time and k is a growth rate constant which we must determine to solve the equation. In 1969 Neil Armstrong became the first man to walk on the moon. In 2009 the European Space Agency launched the Herschel and Planck telescopes that will see back to near the beginning of the universe. 2009-1969 is 40 years. This allows us to write:

$$12\% = e^{k(40)}$$

log 12 = 40k log 2.718

0.026979531 = 0.4342 k

k=0.0621

We now can write:

$$x(t) = e^{(0.0621)t}$$

$$100\% = e^{(0.0621)t}$$

log 100 = (0.0621) t log e

t = 74 years

1969 + 74 years = 2043

Our reasoning would indicate that we will have hyperdrive in the year 2043.

Study summary:

1. We have a 70% chance of developing hyperdrive without destroying ourselves first.
2. We are 12% along the way in development of hyperdrive.
3. We will have hyperdrive in the year 2043, plus or minus.

Sierra Waters was handed the newly discovered document in 2042.

Arthur C. Clarke and Cosmic Archaeology

Monolith

Arthur C. Clarke's monolith turned out to be a computer put on Earth by extraterrestrials and The Moon to give us an evolutionary nudge when we needed it and to monitor us. It had the dimensions of 1 by 4 by 9, the squares of 1, 2, and 3. 1+4 = 5. The height is 9. I have found that 9/5 (which is 1.8) occurs throughout nature in the areas held most sacred to man down through history, the sun, the moon, gold, silver, water and air:

1. If we compare the mass of air to the mass of water and increase that by a factor of the human body temperature to the freezing temperature of water, we get a value that is 9 compared to 5, which is 1.8.

2. If we compare the mass of an atom of gold to an atom of silver, it is 9 compared to 5 (comparing their molar masses).

3. If we compare the radius of the sun, that is the distance from its center to its surface, to the distance from the center of the earth to the center of the moon, it is 9 compared to five.

9 compared to 5 is nine fifths (9/5) which is equal to 1.8

Glancing at my data tables and find that if we take the distance of the planet Saturn to the sun as 9 (closest approach), then the distance to the planet Jupiter from the sun is five (closest approach). In fact this way of measuring distances puts the earth exactly at 1 unit from the sun. This is interesting, because Jupiter and Saturn, aside from being the "middle children" of the solar system, planets 5 and 6 of a planetary family of 9 or 10 depending on whether or not you consider the asteroid belt a planet that did not form, and anything found beyond Pluto a planetoid, these planets carry the majority of mass of the solar system, significantly, and thus embody most of the dynamics of its formation.

I have also found that the basis of computers and AI (artificial intelligence), which is doped silicon, has the golden ratio in the means of its components. The golden ratio is recurrent throughout life because of the dynamics it has to offer. Doped is silicon, phosphorus, and boron. These are naturally occurring elements, made by nature, namely forged in stars. If P is phosphorus, B is Boron, and Si is silicon (geometric mean by Si):

$sqrt(P*B)/Si = sqrt(30.97*10.81)/(28.09) = 0.65$

and let us take the harmonic mean between phosphorus and boron and divide it by silicon:

$(2*(30.97*10.81)/(30.97+10.81))/28.09 = 0.57$

Now let us take the arithmetic mean of these two numbers:

(0.65+0.57)/2=0.61

which are the first two digits in the golden ratio.

The golden ratio is 1.618 to three decimal places. Notice that the 2nd and third digits after the decimal are 1 and 8, the two digits in 9/5. The 1 and 6 add up to 7, the average of nine and five, the 6 minus the one is our five, and, the eight plus the one is our nine. So essentially, we have connected the monolith of Arthur C. Clarke with nature, and computers and artificial intelligence, integrated circuits, transistors, diodes (doped silicon), with the monolith.

Conclusion

The Science Fiction of Paul Levinson, Isaac Asimov, and Arthur C. Clarke are tied up in Humans making Humans, whether robots or clones, for the purpose of saving humanity. And my astronomical discoveries as I go through life making them, seem to be tied up in their stories.

The Mystical Elements Of The Fiction-Reality Entanglement

Seeds

Seeing the movie "Pirates" I there saw a ship with the sea god Neptune, trident in hand, mounted to the front. Years later this image seemed to say it meant something pertaining to the Universe and the Planet Neptune. My astronomical research indeed, years later, ended at that planet, with my Neptune equation.

Finally, as I had not made any progress for a while, another image struck me. It was when I saw a book called "Plutonia" in the personal library of Paul Levinson, in a movie about his book, "Behind The Plot to Save Socrates". Its age, and reddish-pink cover reminded me of my book "The Levinson-Asimov-Clarke Phenomenon", as it is a funky cover designed by myself of a rocket traveling in a pink Universe. My Book itself is, in part, about Paul Levinson's "Plot To Save Socrates". And because Plutonia derives from Pluto, which is a planet, one I had yet to consider in the research for this project, it is about Plutonia.

My book, so far, ends in a phenomenon linking Levinson, Asimov, and Clarke, to the timewave theory of Terence McKenna, which was inspired by his taking hallucinogenic psilocybin mushrooms in a jungle in South America under Shamanistic instruction.

After researching the work Plutonia at wikipedia, I found Plutonia is not just a science fiction novel by a Russian Scientist, written in 1915, but is a type of hallucinogenic psilocybin mushroom called psilocybin plutonia (common name "magic mushrooms").

Since timewave is about interconnectedness, and so is my book, I thought I would post this occurrence here.

Since interconnectedness arises in my research pertaining to Paul Levinson's "The Plot To Save Socrates", and his book is about time travel, I am beginning to understand why: interconnectedness has everything to do with time.

Ian Beardsley
April 17, 2011

April 20, 2011

"Plutonia" arrived in the mail April 20, 2011. I started reading it on that day. The book begins with a team of four scientists leaving Moscow for an Arctic expedition on April 20 (the same date). I am right in phase with their adventure. I also ordered The Silk Code by Paul Levinson on April 20, 2011, realizing after having received an e-mail from Amazon.com that it shipped out on April 20 that this book was in sync with "Plutonia", I ordered The Silk Code because Paul Levinson's "Plot To Save Socrates" was part of the "fiction-reality entanglement" I had been experiencing.

Archimedes Plutonium

Having finished the book "Plutonia" which lead me to consider the last planet "Pluto", among other things already mentioned, I decided to post my work to the discussion group sci.astro, a google group. Something I had postponed for years, and thought maybe would not be a good idea at all do to the given the nature of this work. Immediately posted above my work was the work of an author whose pen name is Archimedes Plutonium. This immediately gave me another aspect of the word Pluto. In his theory, called The Atom Totality Theory, he shows that the element 231 Plutonium describes the myriad aspects of the Universe. Precisely, the numbers of electron shells, and subshells of that isotope of plutonium provide ratios that describe approximations to the value of pi, euler's number e, the background radiation of the universe, the fine structure constant and an explanation for the imaginary number, i.

Furthermore, I found it interesting that the isotope of plutonium that does this is 231, because not only does my work bear a connection to the book "Plutonia", it also bears a connection to 2001: A Space Odyssey and its sequels because the monolith in that book has proportions that are the squares of 1, 2, and 3 which I have explained relate to my discovery of 9/5 throughout nature and 1, 2, 3 are the digits in 231 Plutonium. Once again, the time that I do specific works and read other works, I find my life entangled with literature, no doubt from my trips to Mexico and Italy, whose literature is often based on this phenomenon that is a part of life in those countries. The thing is to find my way out of it, and approach academics from that methodology I practiced while a student at The University of Oregon and working at the state observatory. Perhaps then I can see my work in a new light, one where it is a part of me as opposed to me being a part of it.

Ian Beardsley May 16, 2011

Papers in which I am listed as co-author (Beardsley I.S.)

•

• 1986 _Kemp J. C.; Henson G. D.; Kraus D. J.; Carroll L. C.; Beardsley I. S.; Takagishi K.; Jugaku J.; Matsuoka M.; Leibowitz E. M.; Mazeh T.; Mendelson H._SS 433: A 6 year photometric record_Astrophys. J. 305, 805 (1986)
• Department of Physics, University of Oregon

• 1986 _Kemp J. C.; Henson G. D.; Kraus D. J.; Beardsley I. S.; Carroll L. C.; Duncan D. K._Variable polarization and activity in Arcturus_Astrophys. J. Lett. 301, L35 (1986)
• Department of Physics, University of Oregon

• 1986 _Kemp J. C.; Henson G. D.; Kraus D. J.; Beardsley I. S.; Carroll L. C.; Ake T. B.; Simon T.; Collins G. W._Epsilon Aurigae: Polarization, light curves, and geometry of the 1982–1983 eclipse_Astrophys. J. Lett. 300, L11 (1986)
Department of Physics, University of Oregon, Eugene

FICTION-REALITY ENTANGLEMENT

FEBRUARY 4 2013

BY

IAN BEARDSLEY

Levinson

We began with the closest star to us, Alpha Centauri. It is about 4.2 to 4.3 light years away. We rounded that to four because we are dealing with whole numbers. We then did a random walk, and calculated the probability of landing on Alpha Centauri in 10 random jumps of one light year each. We note that a light year is the distance light travels in one year, a year being the time it takes the earth to travel around the sun, and that light speed is a natural constant. As well we note that Alpha Centauri is a star just like the sun, similar in size, luminosity, and the same spectral class. We rounded our answer to the nearest whole number, and it was that we have a 12% chance of landing on Alpha Centauri.

We then said this was the percent of development towards building hyperdrive, an engine that would take us to the stars. We then said our progress towards hyperdrive may develop according to the natural growth equation:

$(x(t))=(x\_0)e^{\wedge}kt$

To find k we have taken t =40 years, because in 1969 the first man set foot on the moon, and in 2009 we launched the Herschel and Planck telescopes that will see back to near the beginning of the Universe. We have taken x_0 = 1/100 to say that we began with one step out of 100. The result is:

(k) = 0.0621

We have:

(log 100) = (0.0621)t log e

Or, t = 74 years

1969 + 74 =2043

I was reading just before doing the calculation, The Plot To Save Socrates, by Paul Levinson, and his character, his heroine, Sierra Waters, began her adventure in 2042. I call this "fiction-reality entanglement". I wrote in my book: Fiction-Reality Entanglement: Levinson, Asimov, and Clarke:

"In the case of Paul Levinson, scholars in the future travel through time and use cloning, a concept related to artificial intelligence (it is the creating of human replicas as well, but biological, not electronic), and the goal is to save great ancient thinkers from Greece, and to manipulate events in the past for a positive outcome for the future of humanity, just as the robots try to do in the work of Asimov."

As it would turn out, the novelty rating for the year we set foot on the moon in the Sheliak Version of the McKenna Timwave, is our growth rate constant 0.0621 which multiplied by ten and rounded to two places after the decimal it the golden ratio conjugate. The Timewave is supposed to show the way history unvravels, if we take 2012 as the end point.

Perhaps our calculation refers not to hypedrive, but to other issues if we take The Plot To Save Socrates as a data point: It deals with philosophy, values, cloning, and much more. It may even be we are not talking about the Earth, but another civilization on another planet in the Universe. Our actions could be connected to life elsewhere. We just don't know, we only present the calculations.

Asimov

In the case of Isaac Asimov, we are far in the future of humanity. In his Robot Series, Asimov has man making robots whose programming only allows them to do that which is good for humanity. As a result, these robots, artificial intelligence (AI), take actions that propel humanity into settling the Galaxy, in the robot series, and ultimately save humanity after they have settled the Galaxy and made an empire of it (In the Foundation Series).

I now find that Isaac Asimov puts such a development in his science fiction at a similar time in the future, precisely in 2044, only a year after my prediction and two years after Sierra Waters is handed the newly discovered document that initiates her adventure. So, we have my prediction, which is related to the structure of the universe in a mystical way right in between the dates of Levinson and Asimov, their dates only being a year less and a year greater than mine.

Asimov places hyperdrive in the year 2044 in his short story "Evidence" which is part of his science fiction collection of short stories called, "I, Robot".

This is a collection of short stories where Robot Psychologist Dr. Susan Calvin is interviewed by a writer about her experience with the company on earth in the future that first developed sophisticated robots. In this book, the laws of robotics are created and the idea of the positronic brain introduced, and the nature of the impact robots would have on human civilization is explored. Following this collection of stories Asimov wrote three more novels, which comprise the robot series, "The Caves of Steel", "The Naked Sun", and "The Robots of Dawn".

"I, Robot" is Earth in the future just before Humanity settles the more nearby stars. The novels comprising "The Robot Series" are when humanity has colonized the nearby star systems, The Foundation Trilogy, and its prequels and sequels are about the time humanity has spread throughout the entire galaxy and made an Empire of it. All of these books can be taken together as one story, with characters and events in some, occurring in others.
Hyperdrive is invented in I, Robot by a robot called The Brain, owned by the company for which Dr. Susan Calvin works when it is fed the mathematical logistical problems of making hyperdrive, and asked to solve them. It does solve them and it offers the specs on building an interstellar ship, for which two engineers follow in its construction. They are humorously sent across the galaxy by The Brain, not expecting it, and brought back to earth in the ship after they constructed it. This was in the story in "I, Robot" titled "Escape!".

But Dr. Susan Calvin states in the following short story, that I mentioned, "Evidence":

"But that wasn't it, either"…"Oh, eventually, the ship and others like it became government property; the Jump through hyperspace was perfected, and now we actually have human colonies on the planets of some of the nearer stars, but that wasn't it."

"It was what happened to the people here on Earth in last fifty years that really counts."

And, what happened to people on Earth? The answer is in the same story "Evidence" from which that quote is at the beginning. It was when the Regions of the Earth formed The Federation. Dr. Susan Calvin says at the end of the story "Evidence":

"He was a very good mayor; five years later he did become Regional Co-ordinator. And when the Regions of Earth formed their Federation in 2044, he became the first World Co-ordinator."

It is from that statement that I get my date of 2044 as the year Asimov projects for hyperdrive.

Clarke

Everything In 2001: A Space Odyssey by Arthur C. Clarke, Derived From Nine Fifths

In 2001: A Space Odyssey, the ship's computer was called HAL. It was the brain and central nervous system of the ship. It was asked in the movie: Is HAL intelligent? Is he alive? So far I have explained how nine-fifths occurs in the Solar System and Nature. I have also found that a computer like HAL is alive, in a way. HAL was silicon based and I have found that silicon has the golden ratio conjugate in it, which is recurrent throughout life and that nine-fifths is in the golden ratio conjugate. I wrote:

I have also found that the basis of computers and AI (artificial intelligence), which is doped silicon, has the golden ratio in the means of its components. The golden ratio is recurrent throughout life because of the dynamics it has to offer. Doped is silicon, phosphorus, and boron. These are naturally occurring elements, made by nature, namely forged in stars. If P is phosphorus, B is Boron, and Si is silicon (geometric mean by Si):

$$\sqrt{P*B}/Si = \sqrt{30.97*10.81}/(28.09) = 0.65$$

and let us take the harmonic mean between phosphorus and boron and divide it by silicon:

$$(2*(30.97*10.81)/(30.97+10.81))/28.09 = 0.57$$

Now let us take the arithmetic mean of these two numbers:

(0.65+0.57)/2=0.61

which are the first two digits in the golden ratio conjugate.

I have found nine-fifths is in the proportions of the monolith, it dimensions are 1 by 4 by 9. 1+4 = 5. I have talked about how my discovery of nine-fifths in the universe comes from a gypsy shaman via an AE35 antenna of his design, and how we can derive from nine-fifths, a planet composed of europium, and Europa was the important moon in the Space Odyssey Novels of Clarke that became a planet to Lucipher. I shown that HAL is connected to nine-fifths, because it is related to the golden ratio conjugate which is in Silicon. With that, I have shown all of the key aspects of 2001 can be derived from nine-fifths. I wrote:

"The golden ratio conjugate is 0.618 to three decimal places. Notice that the 2nd and third digits after the decimal are 1 and 8, the two digits in 9/5 divided out. The 1 and 6 add up to 7, the average of nine and five, the 6 minus the one is our five, and, the eight plus the one is our nine. So essentially, we have connected the monolith of Arthur C. Clarke with nature, and computers and artificial intelligence, integrated circuits, transistors, diodes (doped silicon), with the monolith."

AE-35 Is Related To Nine-Fifths

"I wrote a short story last night, called Gypsy Shamanism and the Universe about the AE-35 unit, which is the unit in the movie and book 2001: A Space Odyssey that HAL reports will fail and discontinue communication to Earth. I decided to read the passage dealing with the event in 2001 and HAL, the ship computer, reports it will fail in within 72 hours. Strange, because Venus is the source of 7.2 in my Neptune equation which comes from the nine-fifths discovery, and represents failure, where Mars represents success."

We have shown everything in Clarke's Space Odyssey Novels can be derived from nine-fifths.

Levinson, Asimov, Clarke, And The Neptune Equation

I wrote a book where I make a prediction for the invention of hyperdrive. It came out sandwiched between the date Isaac Asimov predicts for hyperdrive (2044) in his science fiction stories, "I, Robot", and the date in which Paul Levinson's Herioin, Sierra Waters begins her adventures (2042), in his book, "The Plot To Save Socrates". The date I predicted with mathematics, was 2043.

I happened to make my predictions during a time when I was reading these works of Levinson and Asimov. You can imagine my surprise when I found Fiction Entangled With Reality, There is much more to this story, and it resulted in my book Fiction-Reality Entanglement: Levinson, Asimov, And Clarke. Many interesting entanglements cropped up.

Just what was the research I was doing when I experienced this entanglement? I was working on my enigmatic discovery of nine-fifths in the Universe, which lead to my book "Are Humans Alone In The Universe And Are They Here For A Reason".

Now I find Fiction-Reality Entanglement: Levinson, Asimov, And Clarke" is entangled with "Are Humans Alone In The Universe And Are They Here For A Reason?" I simply took Arthur C. Clarke's "2061: Odyssey Three" because it was the other book I was reading when I was doing my research and was part of the Fiction-Reality Entanglement: Levinson, Asimov, And Clarke".

I simply thought since Clarke's Odyssey Three, because it takes place in 2061 and that is the time closest to the 2042 of Sierra Waters, the 2044 of Asimov, and the 2043 of my prediction, I would subtract my 2043 from the 2061 of Clarke's Odyssey Three and the result is:

$$2061-2043=18$$

The answer is it takes place 18 years after my prediction of 2043. Divide 18 by 10, the number of fingers we have to count on, and you get one point eight. If you divide the 9 by 5 in my enigmatic nine fifths, you get one point eight, as well. I consider Levinson and Asimov's dates to be upper and lower limits for hyperdrive, because my value is sandwiched between the two. That is it will at the earliest happen in 2042 or by the latest 2044. Either that, or the three dates are speaking of an epoch connected to the stars, in particular Alpha Centauri, because I used it in my calculation, not to mention the speed of light and the time it takes earth to make one revolution around the sun (1 year).

Asimov, Clarke, and Heinlein are considered the Three Pillars of Modern Science Fiction. I see Heinlein more as using science fiction to explore politics. I like to put Levinson, Asimov, and Clarke together as the heart of Science Fiction, which is the spirit of science today, and perhaps Heinlein as the mind, sometimes, to use the Hindu model where The Heart is very important, but means nothing without the guidance of the Mind. Of course all the above authors embark, ultimately in the realm of the mind and the heart.

The Author

www.ingramcontent.com/pod-product-compliance
Lightning Source LLC
Chambersburg PA
CBHW032016170526
45157CB00002B/723